Istanbul (Turkey)

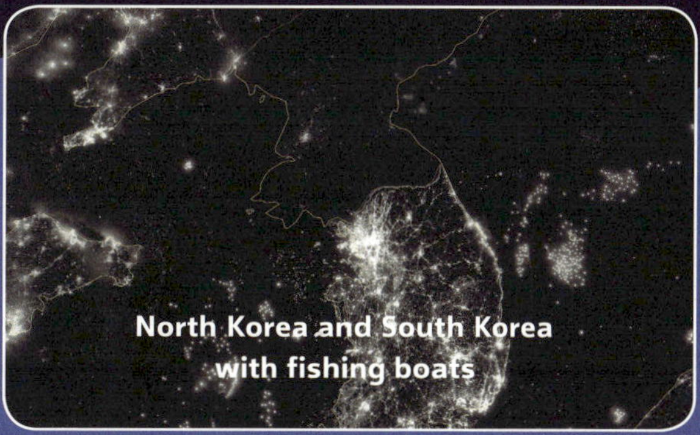
North Korea and South Korea with fishing boats

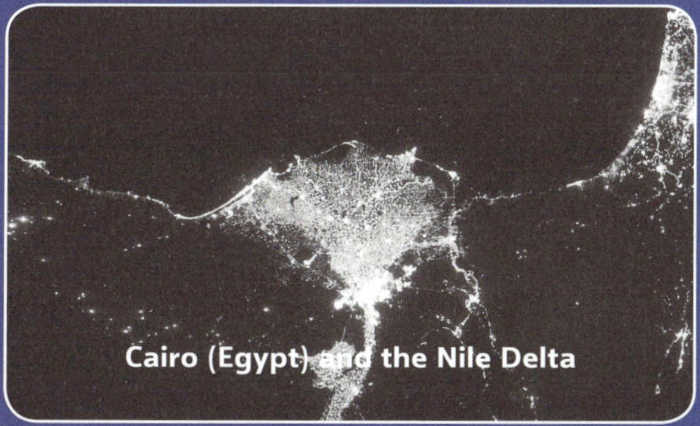
Cairo (Egypt) and the Nile Delta

Bushfires in Australia

Diercke Geography

For Bilingual Classes

Volume 2

Moderation:
Prof. Dr. Reinhard Hoffmann

Autorinnen und Autoren:
Sarah Flöck
Henriette Heß
Reinhard Hoffmann
Verena Hundertmark
Marjana Mehltretter
Dirk Reischauer
Ingo Warken
unter Mitwirkung der Verlagsredaktion

westermann

Cover: Manhattan, New York (USA)

Mit Beiträgen von:
Brigitte Dreymüller, Dieter Haupt,
Volker Friedrich, Dimo M. Rischke

CARL DIERCKE (geb. in Kyritz, Landkreis Ostprignitz/Preußen) lebte von 1842 bis 1913 und war Pädagoge und Kartograph.
Von ihm stammt der bekannte *Diercke-Schulatlas*, der erstmals 1883 unter dem Namen „*Schul-Atlas über alle Teile der Erde*" erschien.

Weitere Informationen finden sich im Internet: *www.diercke.de – (Info/Chronik)*

westermann GRUPPE

Druck A[4] / Jahr 2020
Alle Drucke der Serie A sind im Unterricht parallel verwendbar.

Redaktion: Lars Büttner
Bildredaktion: Susanne Guse
Umschlaggestaltung: Thomas Schröder
Druck und Bindung: Westermann Druck GmbH, Braunschweig

ISBN 978-3-14-**114038**-5

Zur Arbeit mit diesem Buch

Rubriken

Information:
Hier werden Begriffe oder Zusammenhänge kurz erläutert.

FACT SHEET - Steckbrief:
Faktensammlung zu einer Region.

KEY TERMS - Fachbegriffe:
Diese Begriffe solltest du kennen! Für sie wurde daher am Ende des Buches ein Lexikon (Glossary) eingerichtet.

STEPS - Arbeitsschritte:
Hier sind die Arbeitsschritte für eine Methode übersichtlich aufgelistet.

TASKS - Aufgaben:
Die Aufgaben helfen dir, das Gelernte zu üben, zu vertiefen oder zu erweitern. For experts: Diese Aufgaben sind schwieriger und erfordern einen höheren Arbeitsaufwand.

Verweise
- auf das Internet:
- auf Methodenseiten (Geo Skills):
- auf Methodenseiten im TOOLKIT: (ISBN: 978-3-14-**114042**-2)
- auf Themenseiten: (› pp 14-15)
- auf Materialien: (**M**+Nummer)

Atlashinweise
Schlage den Atlas auf der angegebenen Seite auf oder gib den Karten-Code im Suchfeld der Internet-Adresse ein:

Blaues Symbol: 100800-240 schueler.diercke.de
Diercke Weltatlas (2015)

Gelbes Symbol: DE-172 www.diercke.com
Diercke International Atlas

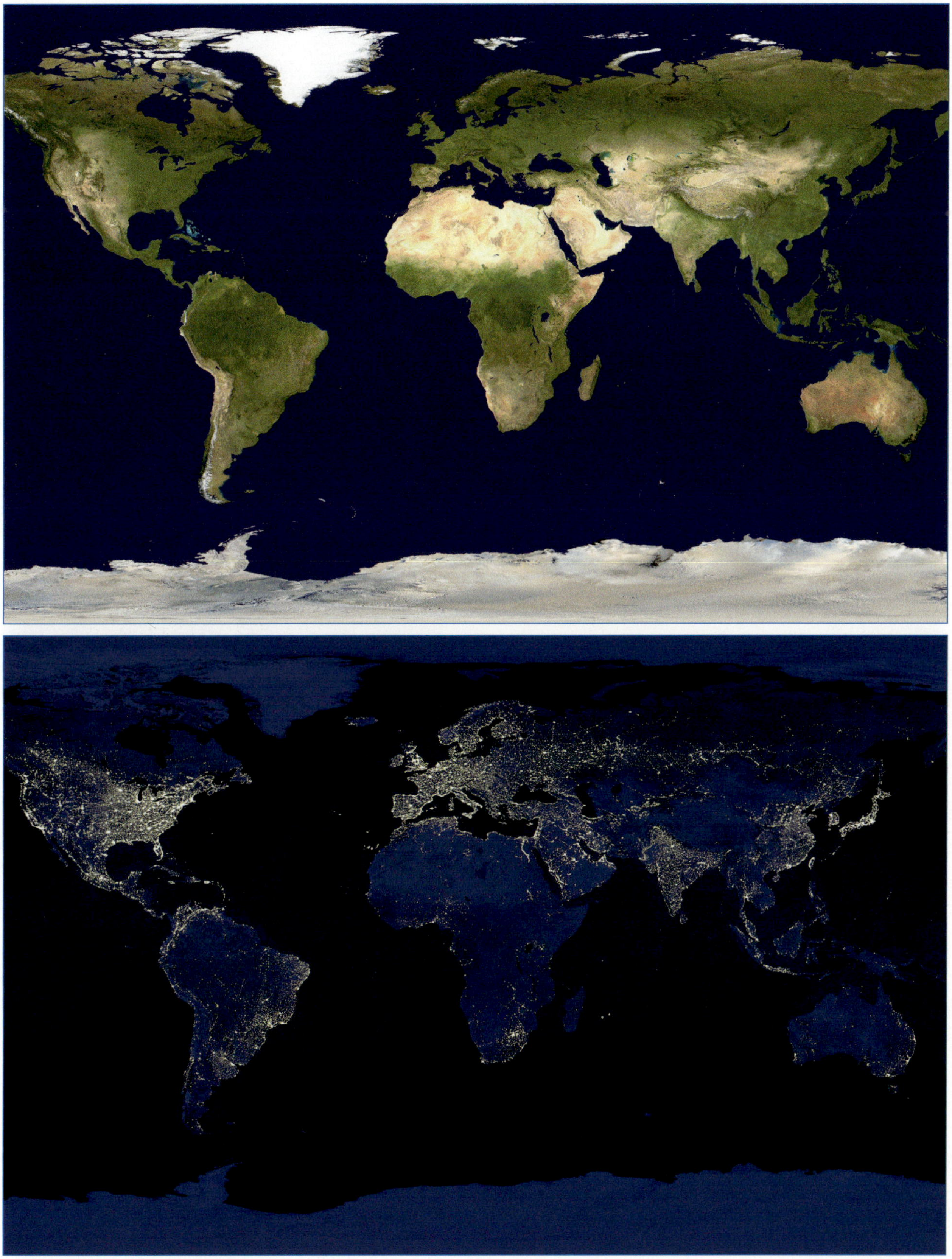

M1 Satellite images of the Earth

Contents

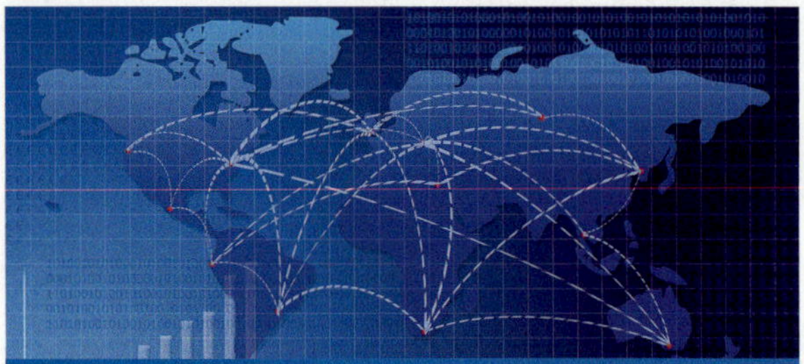

6 Globalisation

8 Globalisation – What Does It Mean?
10 Globalisation or 'The World is Shrinking'
12 Transport in a 'Shrinking World'
14 Information and Communication Technology Changes Life
16 Global Production
18 Location Factors
20 World Trade

22 Global Disparities

24 Globalisation – Worldwide Differences
26 New York – A Global City
28 Singapore – A Transport, Financial and High-Tech Hub
30 Detroit – Industrial Decline in MEDCs
32 Lake Victoria – Once the Rich Fishing Grounds
34 Shanghai – A Fast Changing City
36 Level of Development and Indicators
38 Globalisation – Different Views

40 People

42 World Population
44 The Demographic Transition Model
46 The Challenges of Population Change
48 Feeding the World
50 International Migration
52 Internal Migration

54 Energy Resources

56 Energy – A Vital Component of Our Life
58 Coal – The Dirty but Indispensable Fuel
60 Crude Oil – Still the Black Gold
62 Natural Gas – A Clean Fossil Fuel?
64 Nuclear Energy – Friend or Foe?
66 Hydropower – Electricity from Water
68 Bioenergy – Not Just Firewood
70 Energy – Blown by the Wind
72 Solar Energy – Let the Sunshine In
74 Other Renewable Energies
76 World Energy – Challenges and Perspectives

78 Environment – Global Change

80 Changing Climate
82 People Leave Their Carbon Footprint
84 Soil Degradation – Losing Arable Land
86 Fresh Water – A Vital Resource
88 Threatened Oceans
90 The Ecological Footprint

92 Sustainable Development

94 Sustainability and Sustainable Development
96 Great Plains (USA) – Sustainability Through Soil Conservation
98 Mumbai (India) –Sustainability of Plastic Recycling
100 Indonesia – Sustainable Population Development
102 Think Globally – Act Locally

104 Europe Changes

106 The European Union (EU) and Me
108 The European Union and the World
110 Regional Disparities in the EU
112 The Regional Policy of the EU
114 The Common Agricultural Policy of the EU
116 Connecting Europe – Case Examples
118 The European Car Industry Moves East
120 The EU – Quo Vadis?

122 Germany in a Changing World

124 Germany – A Leading World Economy
126 Demographic Change
128 A Land of Contrasts – Regional Disparities in Germany
130 Regional Migration in Germany
132 Migration to and from Germany
134 A Bright Future? – Germany's Energy Transition
136 Structural Changes in Agriculture
138 Meat Production South of Oldenburg
140 Rise and Fall of the German Coal Industry
142 Structural Change in the Ruhr Area
144 Metropolitan Region Rhine-Main
146 Munich – A Success Story
148 Shrinkage and Growth in the Halle-Leipzig Region
150 Berlin – The Fast Changing Capital

152 Geo Skills

152 How to Work with an Age-and-Gender Diagram
154 How to Work with Cartoons
156 How to Work with Important Instructions

158 Appendix

158 Knowing What to Do
160 Words (Index)
167 Glossary

Quayside crane
TERMINAL BURCHARDKAI
Towboat
Quay dock

Globalisation

8 Globalisation – What Does It Mean?

10 Globalisation or 'The World is Shrinking'

12 Transport in a 'Shrinking World'

14 Information and Communication Technology Changes Life

16 Global Production

18 Location Factors

20 World Trade

M1 Port of Hamburg – gate to the world

Globalisation – What Does It Mean?

M1 Container terminals and container vessels in Hamburg (Germany)

M2 International business meeting via videochat

'Globalisation is the process of increasing interconnection of different national and regional economies, societies, and cultures. Its striking elements are global networks of trade, communication, transportation, and migration.'

'Globalization is not a monolithic force but an evolving set of consequences – some good, some bad and some unintended.'

M3 Entry of Hong Kong Disneyland (China)

M4 Check-in and flight information at Singapore Airport (S.A.)

TASKS

1. a) List at least five products which you use or consume during one day.
   b) Find out where they come from.
2. Explain which aspects of globalisation can be found in the pictures M1-M4.

100800-282
schueler.diercke.de

pp. 12-13
Photographs

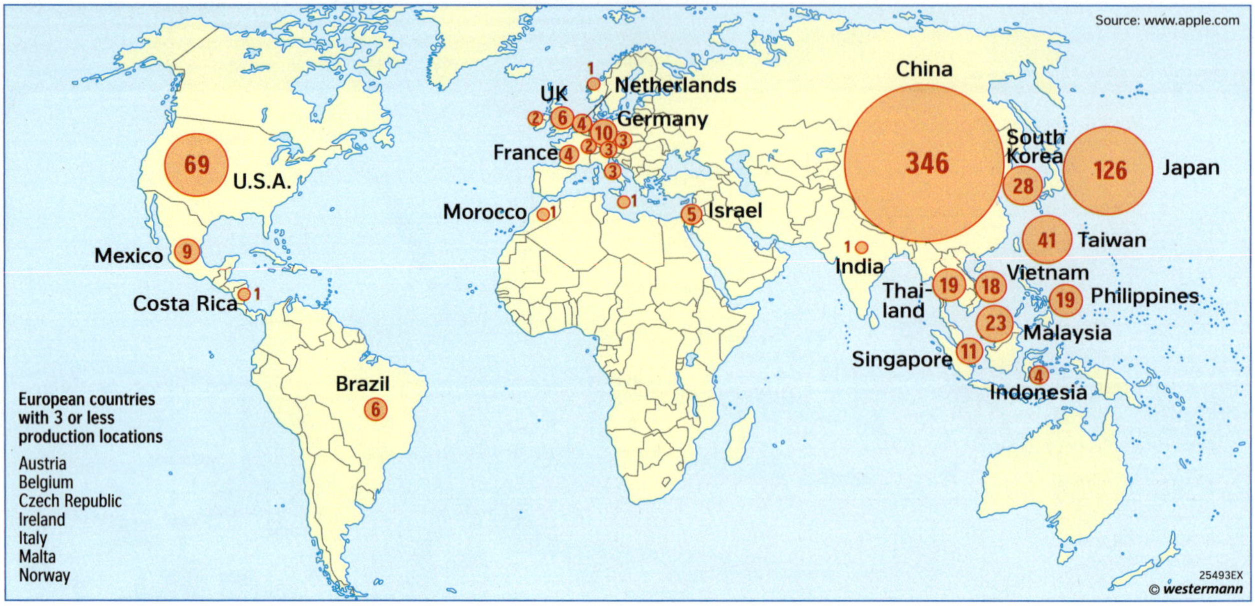

M1 Number of production locations and suppliers of Apple products (2015)

Globalisation or 'The World is Shrinking'

100800-268, -270 schueler.diercke.de

DE-194 www.diercke.com

Internationalisation of economies

There are many aspects to globalisation. One of them is the increasing internationalisation of economies. This means that the production of goods and their sale is not limited to countries and their national markets anymore.

Manufacturing goods has become an international process in which labour is divided between many different locations all around the world (**M1**). Thus, Dutch, German, or Japanese products are more and more turned into 'global products'. Today, goods and capital flow rapidly in a worldwide market.

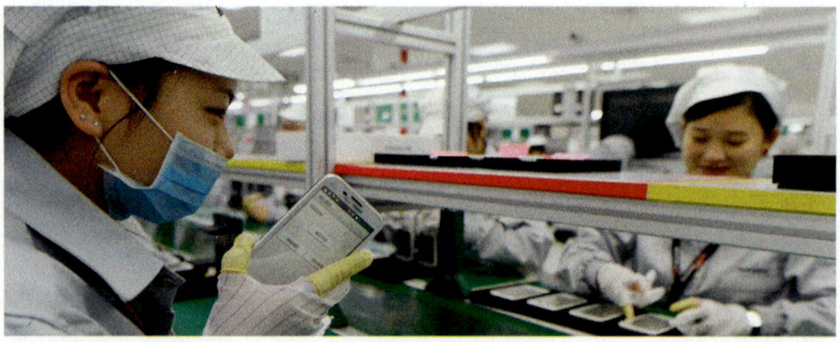

M2 Smartphone assembly line in Wuhan (China)

New dimensions of an old process

The process of globalisation is not a new one. For thousands of years, people have bought and sold products from all over the world. The main differences to the past are the volume and speed of the flow of goods, services, capital, and information around the globe. Today, the volume of world trade is more than twenty times as big as it was in the 1950s.

Additionally, the main actors and driving forces of globalisation have changed fundamentally. Faster transport and communication systems are the two main means by which the world is shrinking (› pp. 12-15).

However, the process of globalisation and its consequences for people, regions, and states are much-discussed topics all around the world – and there are many different opinions on these topics (› pp. 38). Those who economically profit from globalisation have to face critics who focus on the exploitation of workers or the environment, for example.

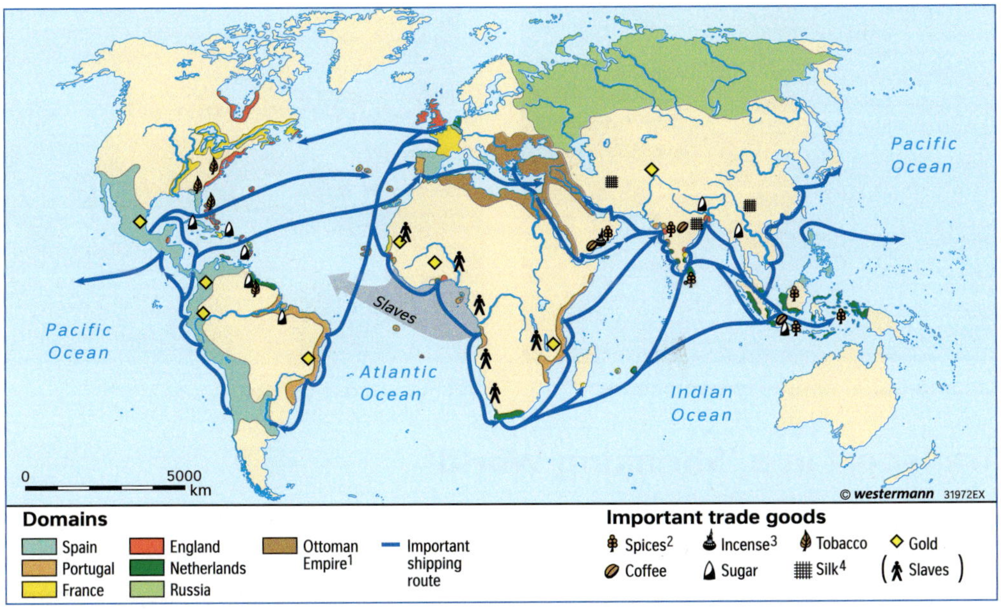

M3 The age of European discovery, expansion, and colonialism (1700)

pp. 10 - 11 Thematic Maps

HELPFUL WORDS

for M3:
1 – Osmanisches Reich
2 – Gewürze
3 – Weihrauch
4 – Seide

for M4:
1 – Zeitalter
2 – Akteur, Handelnder
3 – angeführt

Three Eras [1] of Globalisation

There have been three great eras of globalisation: The first era I called Globalisation 1.0. It lasted from 1492 until the early nineteen hundreds. It shrunk the world from a size large to a size medium. And that era of globalisation was really defined and characterised by countries globalising. You went global through your country. Whether it was Spain exploring the New World, Britain colonizing India, Portugal [colonizing] East Asia. The dynamic agent [2] of globalisation was the country.
Globalisation 2.0 lasted from the early 1820s until the year 2000. It shrunk the world from size medium to size small. Only that era of globalisation was really defined and spearheaded [3] by companies globalising. You went global through your company, companies searching for new markets and for labour.

While you were sleeping, or at least while I was sleeping, we entered Globalisation 3.0, from the year 2000 to the present. It is shrinking the world from size small to size tiny. What is new is that this era of globalisation is built around individuals. So it has really gone from a globalisation built around countries to one built around companies, to one built around individuals.

Source: 'The World is Flat' (Thomas L. Friedman, 2005, abridged)

M4 Extract from a lecture by T.L. Friedman

TASKS

1. Describe where the different production locations and suppliers of Apple are located (**M1**).
2. Sum up the 'first era of globalisation' and its main features (**M3**, **M4**).
3. For experts: Explain in which ways today's globalisation is 'built around individuals' (**M4**).

KEY TERMS

- economy
- globalisation
- labour
- market
- world trade

M1 The modernised Panama Canal connects two oceans

M4 Low-cost carrier

Transport in a 'Shrinking World'

100800-268, 272
schueler.diercke.de

DE-194
www.diercke.com

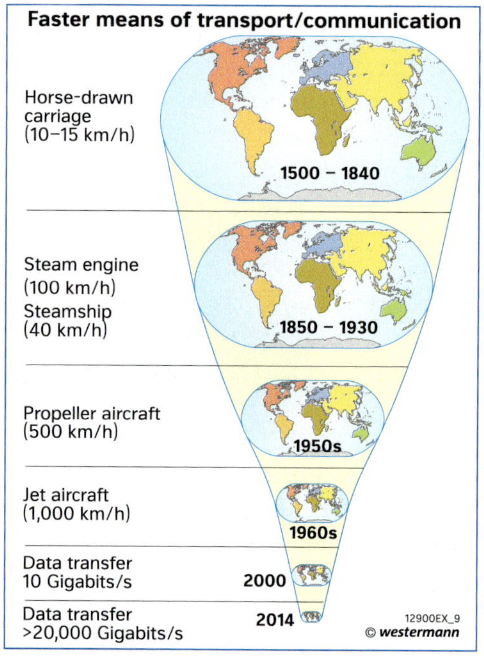

M2 Shrinking transport times

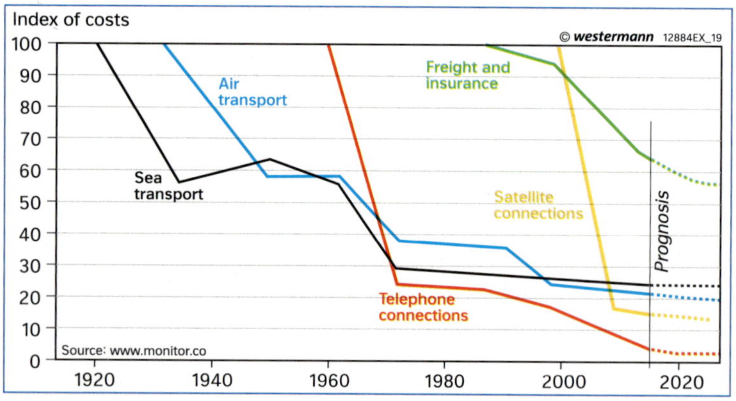

M3 Changes in transport and communication costs

The globalisation of production and the growth of international trade became possible by progress in international transportation. The opening of the Suez Canal in 1869, for example, made it possible to shorten a journey from Great Britain to India on average from six months to two months. Additionally, technologic improvement leads to falling transport costs, which is why the quantities of freight and passengers can rise steeply. Moreover, telecommunication technology accelerates globalisation because communication is no longer linked to transport in many cases.

Important modes of transport

About 50 per cent of world trade takes place between regions that are more than 3,000 km apart. Therefore, two modes of transportation are of greatest importance: maritime and air transport.
The invention of the standardised shipping container and the rise of low-cost air traffic make it easier and cheaper to transport goods or to reach a destination (**M1**, **M4**).

Global container traffic

In 1956, the first container vessel was used to transport goods from New Jersey to Houston, Texas (USA). Today, there are about 30 million single containers criss-crossing the oceans. Between 1996 and today, the number of container vessels has increased from 1,700 to more than 5,100.

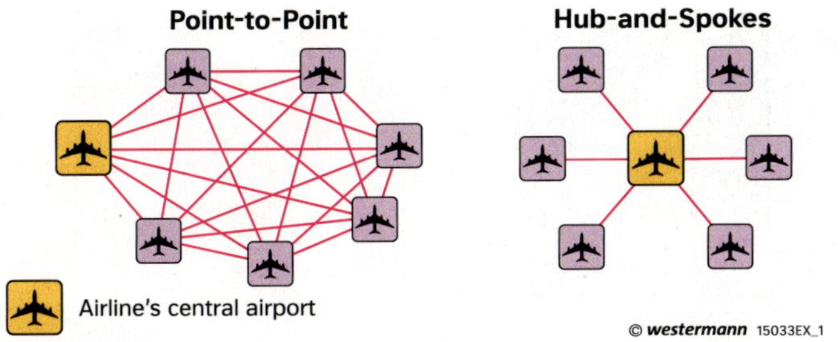

M5 Air transport system models

Twenty Foot Equivalent Unit (TEU)
Cargo capacity of a standard shipping container; most common dimensions: 20 feet long x 8 feet wide x 8.5 feet high (6.10 metres x 2.44 metres x 2.59 metres).

At the same time, the transport capacity of these ships has become larger. Shipping companies use vessels that can carry up to 19,000 TEU. Some of the vessels are more than 400 m long. Consequently, the costs of container transport are declining.
However, the increasing size of container vessels brings new problems. Many new vessels are too large for the Panama Canal, and thus the waterway had to be widened and deepened (**M1**).

'Now everyone can fly'

Air traffic is marked by impressive growth rates, and air travel has become a booming branch of the economy. Most airlines are improving their infrastructure in order to transport larger quantities of freight. Transported goods reach from Korean high-tech equipment to roses from Kenya. Competition between airlines is increasing. This can be seen in the growing number of low-cost carriers, for example, which offer only basic but cheap service. Nevertheless, air transport is concentrated on a few regions of the world.

Today, airlines use different concepts to organize their transportation systems. Most airlines have a limited number of central airports (hubs), where passengers change planes to reach their destination.

HELPFUL WORDS AND PHRASES
for TASK 3:
- The point-to-point system ...
  ... is marked by ...
  ... differs from ...
  ... can be characterised by ...
- In contrast to ...
- While ...
- Compared to ...
- In comparison to ...
- The main difference between ...
- Not only ... but also ...

International Airport Source: ACI (2015)	Freight in million t (world rank) 2005	2015
Hong Kong	3.4 (2)	4.4 (1)
Memphis	3.6 (1)	4.3 (2)
Shanghai	1.9 (9)	3.3 (3)
Anchorage	2.6 (3)	2.6 (4)
Incheon	2.2 (5)	2.6 (5)
Frankfurt	1.9 (7)	2.1 (9)

M6 Leading airports (cargo)

International Airport Source: ACI (2015)	Passengers in million (world rank) 2005	2015
Atlanta	85.9 (1)	101.5 (1)
Beijing	41.0 (15)	89.9 (2)
Dubai	23.6 (40)	78.0 (3)
Chicago	76.5 (2)	76.9 (4)
Tokyo	63.2 (3)	75.3 (5)
Frankfurt	52.2 (8)	61.0 (12)

M7 Leading airports (passengers)

TASKS

1. Describe the development of transport and communication costs between 1920 and today.
2. Give reasons why the world has been shrinking in the past 500 years.
3. Compare the two different air transport models (**M5**).
4. Calculate the time saved by the Panama Canal for a container vessel that moves from Havana to Seattle at a speed of about 40 km/h. Use your atlas.

KEY TERMS
- container vessel
- freight
- hub-and-spokes system
- passenger
- point-to-point system

M1 Information and communication technology connects people around the world

Information and Communication Technology Changes Life

100800-282-02 schueler.diercke.de

DE-194-02 www.diercke.com

pp. 10 - 11 *Thematic Maps*

pp. 16 - 17 *Tables*

Major technological innovations

Advances in information technology have not only transformed the economy but also changed the lives of people all over the world.

When the first long-distance telephone line was completed in California in 1877, it covered a distance of about 97 kilometres. Since then, innovations such as the introduction of communication satellites, the use of fibre optic cables, and constant advances in computer technology have led to a digital revolution. Two innovations which have spread massively since the late 1980s are particularly important in this context: the advent of electronic mail and the world wide web (WWW) based on the internet, and cell phone services.

Vast differences in internet access rates have created a digital divide between more economically developed countries (MEDCs) and less economically developed countries (LEDCs, › pp. 36-37).

However, modern information and communication technologies have also opened up new opportunities for people outside the industrialised world (**M4**).

Country	Fixed phone subscriptions (per 100 people)	Mobile phone subscriptions (per 100 people)	Internet access (per 100 people)	Average monthly costs of fixed telephone subscriptions (+ 30 calls, in US-$)	Average monthly costs of mobile phone subscriptions (+ 30 calls and 100 text messages, in US-$)
Burkina Faso	0.4	80.6	9.7	14.2	11.7
South Africa	7.7	159.3	52.6	19.5	20.4
India	2.0	78.8	36.5	2.5	2.9
China	16.5	93.2	52.3	4.8	4.0
South Korea	58.1	118.5	92.1	6.2	18.6
Venezuela	24.8	92.9	61.5	1.2	10.6
Germany	54.9	116.7	88.4	27.4	18.0

Source: World Bank, Internet World Statistics (2015)

M2 Access to modern telecommunication

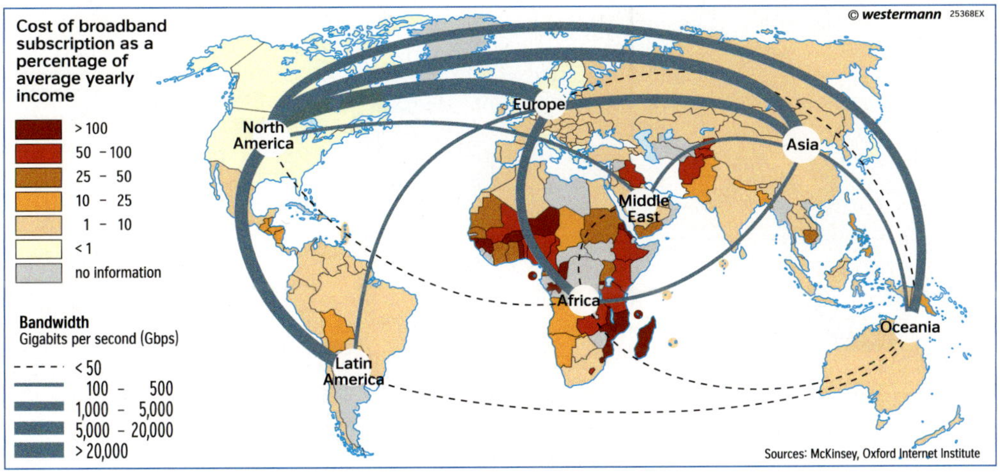

M3 Global digital divide – broadband affordability and data flows (2014)

HELPFUL WORDS AND PHRASES
for TASK 1:
- Without these technologies ...
  ... I would / wouldn't ...
  ... it would be more difficult / harder to ...
  ... it wouldn't be possible to ...
- These technologies ...
  ... help me to ...
  ... make it easier / less difficult to ...

Computer-based jobs in India

Young Indian people sit in small cabins. Their eyes are fixed on their computer screens. They have the entire English-speaking world on the other end of the line, around the clock.
Morning is the best time to reach Australia, then Europe wakes up, followed by the East Coast of the United States and then the West Coast. The flood of calls never stops ...

... "Yes, Sir, of course you can replace your lost credit card in San Francisco today. Just go to ..."
... "Hello, Kansas! So your PC is misbehaving? Let's see what we can do ..."
... "No, Madam, I cannot come to your flat in London's West End but I would be happy to help you fill in your insurance forms over the phone ..."

Source: Spiegel Special International Ed., 'From Poorhouse to Powerhouse' (07/2005)

Call Centres in South Africa

In the recent past, South Africa attempted to attract outsourced businesses from the English-speaking world. Relatively high costs in comparison to India or the Philippines, as well as a comparatively poor country reputation have been preventing widespread growth in this segment.
However, there are important industries mainly located around South Africa's major cities Johannesburg, Durban, and Cape Town. It is highly likely that South Africa will see the number of its clients rise in the near future.
In other parts of Africa the situation is similar but a bit more positive. There are, for instance, some small Sub-Saharan success stories to be found: Kenya and Ghana have successfully expanded their call centre industries. Although the number of people employed in this branch is still not vast, steady progress is made.

M4 India vs. Africa – call centre services in LEDCs

TASKS

1. Name ways in which your life would be different without modern information and communication technology.
2. Sum up the data presented in **M3**.
3. Explain the term 'digital divide' with the help of **M2** and **M3**.
4. Discuss chances and problems of establishing call centre services across the African continent in English or German.

KEY TERMS

- digital divide
- internet
- LEDC
- MEDC
- WWW

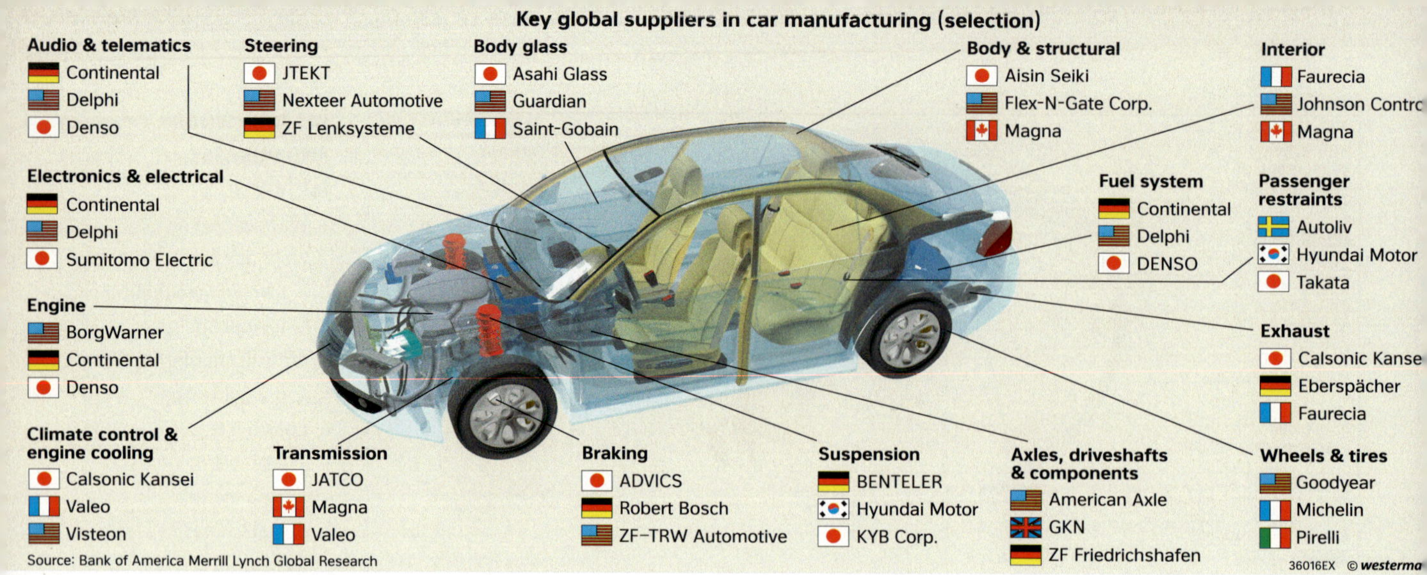

M1 Who makes the car?

Global Production

Globalisation has changed the way goods are produced. Manufacturing commodities has become a global process, which is characterised by the division of labour between different locations around the world (**M1**, **M2**). Companies also buy goods and services from anywhere in the world which is called global sourcing. In addition, production steps can be outsourced to external suppliers.

Mr. Stevenson, it is often argued that globalisation has changed the world economy and the production of goods. Do you agree?

'Oh yes, I do! In fact, it has become so much easier to buy and sell products, services, components and, especially for us in the automotive industry, production processes from and to the entire world. This is of course possible today because of globalisation processes and developments in transport and communication.'

So, this works because there are better phone networks and faster cars?

'No, not that. But faster transport and high-speed internet connections have an impact on businesses and markets. It is much cheaper to transport goods and to communicate with other businesses in comparison to the 20th century, for example. And with sinking costs come sinking prices for goods and services and that is good for all of us.'

But isn't there the risk of competition from other companies all around the globe?

'Yes, there is of course. But there are ways to manage this competition. You have to be better, faster, or less expensive! That is why we do not produce everything near our headquarters. We have other, specialised companies, which work for us. These component suppliers produce at locations all around the world. This is called outsourcing. Of course, there is also competition between these companies but that is good for us because then the quality of the product component rises while the prices are falling.'

I guess this means that you need extremely big production sites because of all the components you need to store?

'Oh no, not at all. Lean production – that is the outsourcing of production steps – does not depend on huge storage facilities. It all depends on transport and communication networks, as I have already mentioned. Product components are delivered just-in-time from global sources. This means that the component suppliers make sure that we get the components we need when we need them in the production process. The key is a good logistics system: they produce and send their component to our assembly site. When it arrives, we unload the truck and use the component directly. This way, none of us needs storage facilities anymore.'

M2 Production trends in the automotive industry

M3 Tunisia – production of cables for the car industry

M5 France – assembly of cars

Made by Bosch

Bosch is an internationally active group of companies with its headquarters near Stuttgart. Bosch is a global player. It has 260 productions sites around the world including more than 60 in Germany. The company has about 300 subcompanies in 60 countries and their products are available in 150 countries worldwide.
Automobile technology adds up to 57 per cent of Bosch's turnover. It produces car electronics, multimedia, steering components, and engine parts for original equipment manufacturers (OEM).
To keep its dominant position among the most important component suppliers, Bosch invests more than 3 billion € per year in research and development. The results are several thousand patents each year. In 2015, Bosch had 375,000 employees and a total turnover of 70.6 billion €, making it the most successful year in the company's history.

M4 Bosch – component supplier and global player

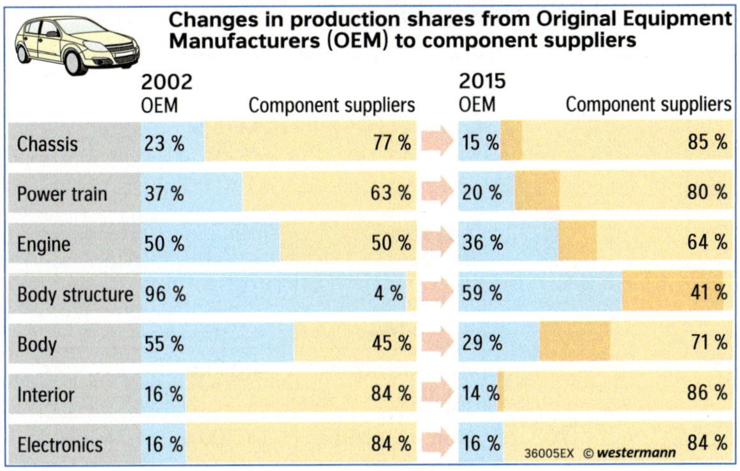

M6 Changes of in-house production depth (vertical integration) in the car industry

M7 Just-in-time?

100800-268
schueler.diercke.de

pp. 20 - 21 Bar Charts

pp. 32 - 33 Texts

KEY TERMS

- component supplier
- global player
- global sourcing
- just-in-time
- lean production
- outsourcing

TASKS

1. Give an overview of the forms of production presented on this page.
2. Describe the changes of vertical integration of OEMs (M6).
3. Explain the roles of OEMs and component suppliers today (M4, M6).
4. For experts: Discuss problems and chances of these forms of production.

M1 Transport infrastructure in Sydney (Australia)

Location Factors

Location factor	Vidin (Bulgaria)	Stuttgart (Germany)
Labour costs (industry)	3.40 € / h	38.99 € / h
Annual income (skilled worker)	4,050 €	37,272 €
Annual income (engineer)	14,700 €	74,331 €
Cost of land for industries	17 € / m^2	149 € / m^2
Electricity costs for industries	69 € / 1,000 kWh	151 € / 1,000 kWh
Water costs	6,960 € / 10,000 m^3	25,900 € / 10,000 m^3
		Data from several sources

M2 Average costs at different locations

Finding the best location

Finding the best location for an enterprise depends on lots of different location factors. Their quality varies from place to place, and differs greatly for different companies and functions of the location. This means that a site for the mass production of clothing has requirements that are very different from those of a management headquarter for an insurance company or new research facilities (**M3**, **M4**). Additionally, the importance of different location factors can change over the course of time.

Because companies look for the best possible location in order to have an advantage over other enterprises of the same branch of business, there is intense

M3 Cheap labour force in Vietnam

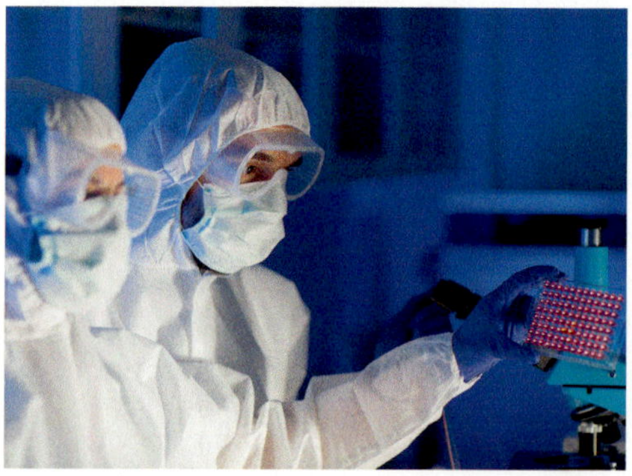

M4 Experts and research facilities (USA)

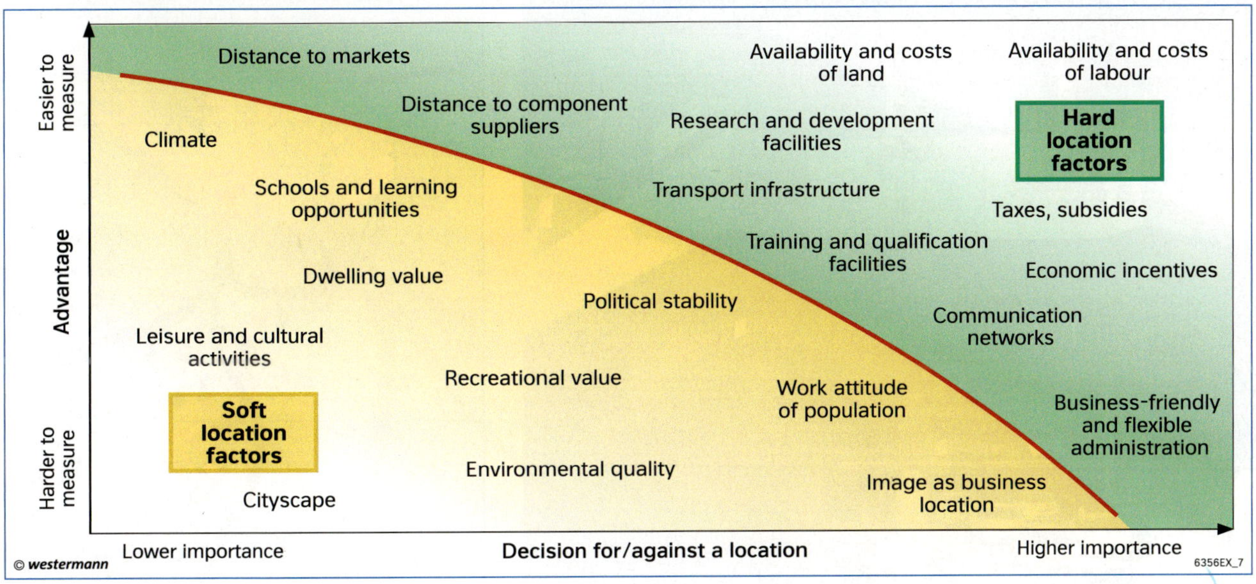

M5 Hard and soft location factors

competition between locations. In a globalised world, regions and countries compete with each other because they need to attract successful, innovative, and future-oriented companies. The economic success of a location and the success of the local companies and their products often depend on each other.

Hard and soft location factors

There are two groups of location factors (**M5**). Hard location factors have a direct impact on the decision for or against a particular location. This group includes factors such as natural resources, the availability of workers, the quality of communication networks, and the distance to markets. Since towns and cities have a great interest in attracting companies, they often offer considerable tax cuts and pay for the improvement of the infrastructure for transport and energy supply.

Soft location factors have an indirect influence on the decision for or against a particular location. This group includes factors such as facilities for education, cultural attractions, the quality of residential areas, a favourable environment, and the image of the region.
While it is easy to calculate the costs for hard location factors such as taxes, transport, or wages, soft location factors cannot be measured monetarily.

HELPFUL WORDS AND PHRASES
for TASK 3:
- On the one hand … on the other hand …
- What it comes down to is that …
- One can say that …
- It is most obvious that …
- Not only … but also …
- … seems to be the most persuasive location factor

TASKS

1. Describe why it is so important for companies to find a good location for their production facilities or offices.
2. Explain why the textile industry of some LEDCs has become competitive in comparison to that of Europe.
3. Discuss which location factors might be the most important for
   a) the mass production of a product,
   b) an office building for the management of a company,
   c) research facilities for a high-tech product.

KEY TERMS

- infrastructure
- labour force
- location factor
- natural resource

Apartment blocks
Recreational area

Shanty huts
THE
KIBERA MEDICAL CENTRE
NURSING & MATERNITY HOME
Open sewer

Global Disparities

24 Globalisation – Worldwide Differences
26 New York – A Global City
28 Singapore – A Transport, Financial and High-Tech Hub
30 Detroit – Industrial Decline in MEDCs
32 Lake Victoria – Once the Rich Fishing Grounds
34 Shanghai – A Fast Changing City
36 Level of Development and Indicators
38 Globalisation – Different Views

M1 Living conditions in Hong Kong and Kibera, a slum of Nairobi

M1 Rich and poor countries – the size of the area equals the GDP of each country (2016)

Globalisation – Worldwide Differences

100800-270-01
schueler.diercke.de

DE-192, DE-194
www.diercke.com

pp. 18 - 19
Line Graphs

pp. 10 - 11
Thematic Maps

Globalisation – only winners?

Globalisation has supported the development of the world's economy. Since 1960, the global GDP has increased from about 1.4 trillion US-$ to about 74,2 trillion US-$ in 2015. However, globalisation is not a process that produces only winners.
Critical voices claim that globalisation brings prosperity to only a few, while worsening the situation of many others. Although globalisation probably decreased poverty in LEDCs, the gap between the world's poorest and richest countries or even parts of the same country has rather been widened (**M4**).

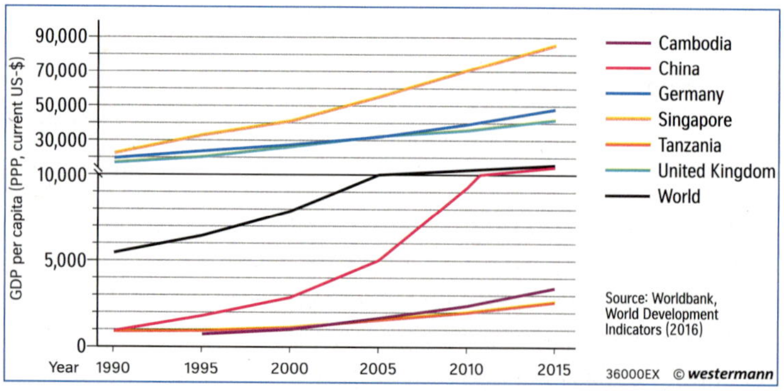

M2 GDP per capita development of several countries

Global fragmentation

One result of globalisation are regional and global disparities. Not all countries or regions benefit from globalisation the same way. At present our world is characterised by a number of islands of prosperity in an ocean of poverty.
Triad economies (**M3**) are the most powerful centres of global economy. The core regions within triad economies contain much of the world's wealth. These prospering regions often specialize in the production of consumer goods, high-tech or providing services. Meanwhile, the NICs of e.g. Asia with their new industries resulting from outsourcing, free trade zones or special economic zones have become a part of that group of prospering regions.
The periphery consists of the regions in the rest of the world that do not profit from global wealth and are rather characterised by weaker economies, poverty and a lower standard of living.
However, in both the core and the periphery, the economic characterisation is not uniform. Poverty can exist in the core, while pockets of economic prosperity can be found in the periphery.

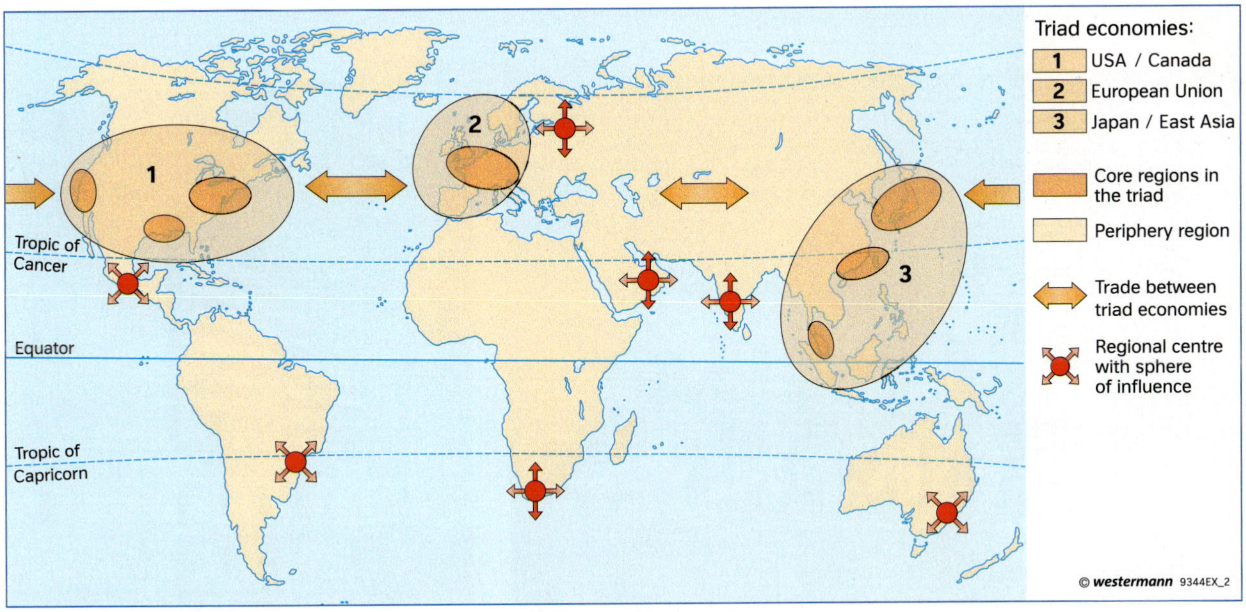

M3 Fragmentation – core and periphery regions

Gross domestic product (GDP)

GDP is the market value of all goods and services produced within a country in a year. It is often used as an indicator of a country's material standard of living. The GDP per capita further indicates the divide between rich and poor nations.

- The 62 richest people own as much as the poorer half of the world population Source: Oxfam (2016)
- The world's richest person: Bill Gates: 76.3 billion US-$ (USA; Microsoft) Source: Forbes (2016)
- 1.2 billion people have to live on less than 1 US-$ a day Source: UN (2016)

M4 Global disparities – facts

The Global Gap

The global inequality crisis is reaching new extremes. The richest one percent now have more wealth than the rest of the world combined. Power and privilege is being used to skew the economic system to increase the gap between the richest and the rest.
A global network of tax havens further enables the richest individuals to hide $7.6 trillion. Meanwhile, the wealth owned by the bottom half of humanity has fallen by a trillion dollars in the past five years. This is just the latest evidence that today we live in a world with levels of inequality we may not have seen for over a century.

Source: www.oxfam.org (Briefing Paper Jan 2016)

M5 Increasing inequality?

TASKS

1 a) Describe differences on Earth with the help of **M1**.
b) Compare **M1** to the GDP development of several countries (**M2**). Use your atlas.

2 Explain the phrase 'island of prosperity in an ocean of poverty' (**M3**, text).

3 Explain two facts of **M4** or **M5** that you consider most remarkable. Answer in English or German.

HELPFUL WORDS AND PHRASES
for TASKS 1 & 2:

- The map shows ...
- A small / large size of a country indicates ...
- In comparison to a map in the atlas, this map ...
- The phrase indicates that ...
- ... marked by prosperity ..., whereas ... marked by ...
- It expresses ...

KEY TERMS

- disparity
- fragmentation
- free trade zone
- gross domestic product (GDP)
- outsourcing
- special economic zone
- triad economy

M1 The towering One World Trade Center in Manhattan is the highest building in the western hemisphere

M4 The New York Stock Exchange – number one in the world

New York – A Global City

Global Disparities

100800-218, -219 schueler.diercke.de

DE-142 www.diercke.com

pp. 12 - 13 *Photographs*

pp. 26 - 27 *Pie Charts*

City	2016 rank	2015 rank
London	1	2
New York	2	1
Paris	3	3
Tokyo	4	4
Hong Kong	5	5

Source: www.atkearney.com (2016)

M2 Ranking of global cities

Stock exchange	Location	Value of shares in trillion US-$
New York Stock Exchange	New York City	19.2
NASDAQ	New York City	6.8
London Stock Exchange	London	6.1
Japan Exchange Group	Tokyo	4.5
Shanghai Stock Exchange	Shanghai	4.0
Hong Kong Exchanges and Clearing	Hong Kong	3.3
Euronext	e.g. Paris, London	3.3
Shenzhen Stock Exchange	Shenzhen	2.3
TMX Group	Toronto	1.9
Deutsche Boerse AG	Frankfurt	1.8

Source: www.worldatlas.com (2015)

M3 Biggest stock exchanges by value of shares

Global cities have emerged in industrialised countries in the process of globalisation since the 1980s. These are cities of great economic, social, and political importance internationally. They are seats of transnational companies, headquarters of international organizations and of financial institutions, homes of cultural institutions and important transport hubs (**M2**).

Economy

New York can be considered to be the world's most important financial center. The world's largest stock exchanges are located in Lower Manhattan (**M3**). Therefore, all important banks, brokerage and insurance companies have their headquarters here. As the leading center of finance in the world economy, the collapse of the Wall Street in 2008 had great impacts on the global economy and led to the worldwide financial crisis.

Moreover, New York City is home of the headquarters of many of the world's largest companies. Of the top 500 companies with the highest turnover worldwide, 17 had their headquarters in New York

M5 Times Square – the world-famous Broadway is here

M6 United Nations General Assembly

(2014). Due to the economic importance of finances, high technology, insurance, and health care, but also branches like the mass media and journalism, over 80 per cent of the employees in New York are employed in the tertiary sector (2014).

Social life and culture

The New York metropolitan area has about 19 million inhabitants, of which about 8 million live in New York City (2015). The ever-increasing population shows New York's continuing national and international importance.

The city is characterized by the wide variety of ethnic groups living there (**M7**). New York is a melting pot for different cultures, though some ethnic segregation exists. Examples include Chinatown and Little Italy.

New York is also of great cultural importance with 500 art galleries, 200 museums, approximately 150 theatres, a world-famous philharmonic orchestra, opera company, and ballet, as well as 20,000 restaurants serving food from all over the world.

Politics

The United Nations is headquartered in New York City. Here, decisions of global political importance are made regularly by the Security Council or by the General Assembly in yearly meetings.

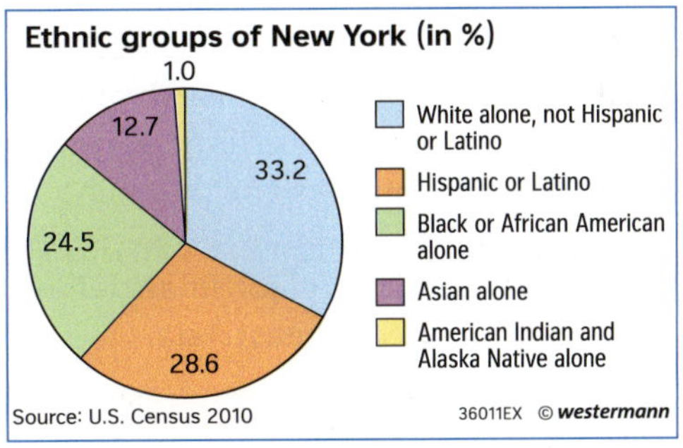

M7 Multicultural New York

TASKS

1. a) Name characteristics of a global city.
   b) Describe them in the photos.
2. Explain why New York is a global city.
3. For experts: Analyse characteristics and reasons for London being global city no. 1 with the help of the internet.

HELPFUL WORDS AND PHRASES
for TASK 2:

- What characterizes New York as a global city is ...
- The fact that ... makes it a global city.
- Usually cities don't have ... while in New York there is / are ...
- Because of the accumulation of institutions like ... New York can be identified as ...

KEY TERMS

- ethnic group
- global city
- segregation
- stock exchange
- transnational company

M1 Port of Singapore

Singapore – A Transport, Financial and High-Tech Hub

100800-186, -193
schueler.diercke.de

DE-097, DE-194
www.diercke.com

pp. 10 - 11 *Thematic Maps*

pp. 18 - 19 *Line Graphs*

pp. 20 - 21 *Bar Charts*

Singapore's booming economy

Singapore has developed from a rather poor country to a nation whose wealth equals those of the richest European countries.

The economic success goes back to the boom of industry, which was driven by favourable location factors, including a cheap labour force, industry-friendly legislation and subsidies. Since then, some of these industries have moved to other Asian countries such as China in search of more favourable conditions. Singapore has maintained its international standing by becoming an important center for import and export and is a major financial and high-tech hub in Southeast Asia.

A favourable location

One reason for the success story is the close location to major international sea routes. About 1,333,000 vessels call at the port of Singapore per year. It is a hub for the largest container vessels and linked to all Asian countries via feeder traffic.

Shipping lines cooperate and form alliances to offer regular routes. Vessels of these alliances follow fixed schedules linking hubs all over the world. Also, Singapore is a major port of call for nearly all Asian routes.

The turnover of containerised cargo has risen in Singapore from about 325 million tonnes per year in 2000 to approximately 576 million tonnes in 2015.

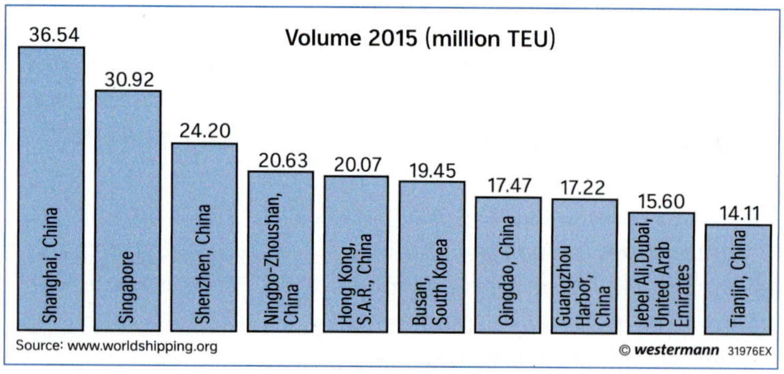

M2 The largest container ports worldwide

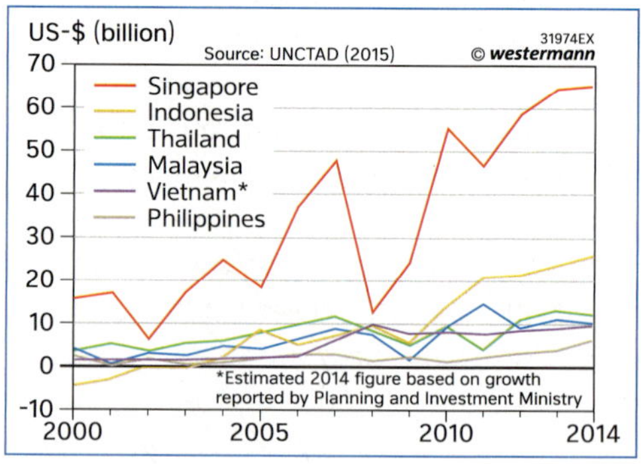

M3 Foreign direct investments in Southeast Asia

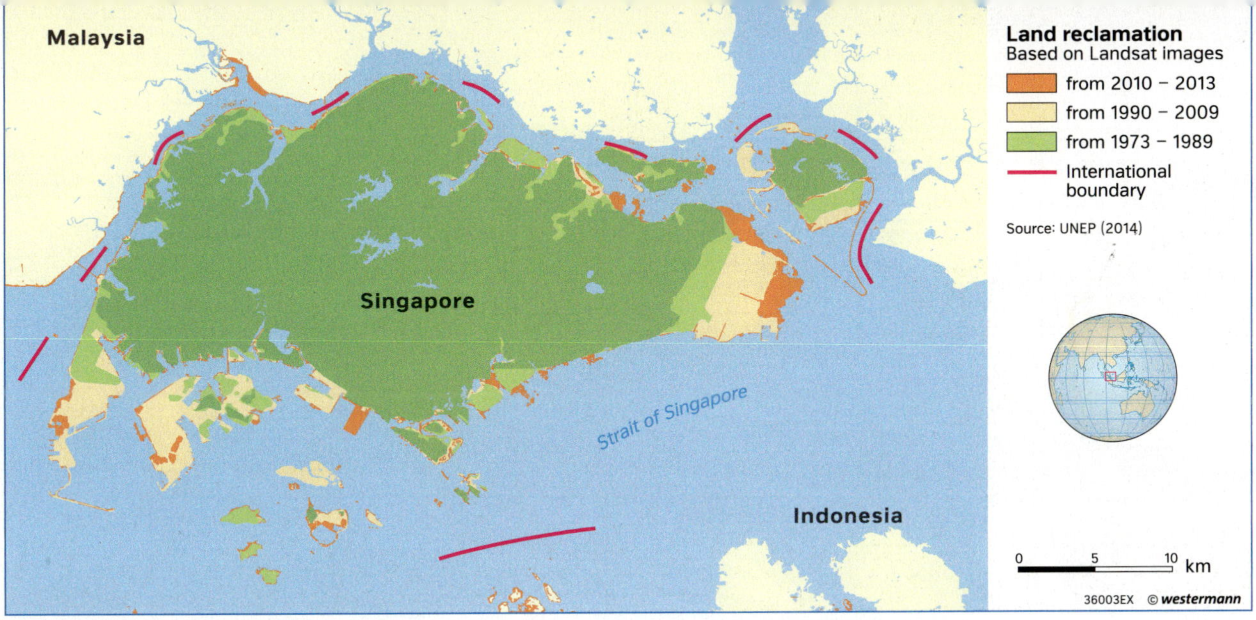

M4 Area development of Singapore

Singapore – a major economy

Singapore has undergone a huge economic development since its independence from the United Kingdom in 1965. The country with limited amount of space and no natural resources has developed into an important manufacturing and financial centre.

This development was possible due to many incentives like low taxes, a liberal immigration policy and state funds. Singapore's global importance especially becomes obvious in the sectors shipbuilding, electronics and (private) banking. Economists see Singapore as a free-market success story.

Today the city is a mature economy. Due to the fact that it had to compete with other low-wage countries, that sometimes emulated its economic policies, Singapore has seen its growth slow down recently.

M5 Singapore – a closer look

FACT SHEET • Singapore

- **Inhabitants:** 5,535,000
- **Area:** 718 sq km
- **GDP:** 292.7 billion US-$
- **Share of economic sectors/GDP:** Primary: 0.0 %; Secondary: 25 %; Tertiary: 75 %
- **Share of employment/economic sectors :** Primary: 0 %; Secondary: 28 %; Tertiary: 72 %
- **Unemployment rate:** 3.3 %
- **Import:** 298 billion US-$
- **Export:** 350 billion US-$
- **Import goods:** electronics, petrochemistry, machinery
- **Export goods:** electronics, petroleum oils, chemicals, accessories of automatic data processing, machinery

M6 Basic information about Singapore (2015)

TASKS

1. Point out the advantages of Singapores location (text, atlas).
2. Explain the importance of the port for Singapore's economy (**M1**, **M2**).
3. Discuss the strengths and limits of Singapore as a transport, finance and high-tech hub (**M3-M6**).

KEY TERMS

- high-tech industry
- hub
- location factor

M1 Detroit – fallow land and Central Business District (CBD)

Detroit – Industrial Decline in MEDCs

The economic boom

Located in the state of Michigan, Detroit experienced an economic boom in the first half of the 20th century. Detroit's ideal transport connection, the availability of cheap raw materials, and Henry Ford's invention of the assembly line (Fordism) led to becoming the world's leading city for car manufacturing. Also, General Motors and Chrysler had their headquarters in Detroit ('The Big Three'). The economic development of Detroit led to growth in the city's size and population (**M2**).

The enduring economic crisis

Starting in the 1960s, Detroit's car industry experienced an economic decline due to the oil crises, increasing foreign competition, and a decrease of productivity.

Between 1960 and 1990, 40 of the 126 auto production sites were closed. These shut-downs caused 40,000 workers to be layed-off. The unemployment and poverty rates grew as the population shrank. As a result, an increasing number of homes became derelict and land fallow. The impact of this decline can still be seen today.

Struggle for revitalisation

Since the city's crisis, especially in Detroits CBD various revitalization projects have been put to practice to make the city more attractive again and to support its new economic development. Since the 1990s, Detroit is focussing on attracting high-tech industries and services. This leads to an increasing segregation of well-paid professionals and badly-paid low-skilled workforce.

Year	Inhabitants
1900	285,704
1920	993,078
1940	1,623,452
1960	1,670,144
1980	1,203,339
2000	951,270
2010	713,777
2015	677,116*

Source: US Census Bureau, *estimate

M2 Detroit – population

Year	Unemployment rate in %	
	Detroit	USA
2000	6.6	4.0
2002	11.8	5.8
2004	14.1	5.5
2006	13.4	4.6
2008	15.0	5.8
2010	24.8	9.6
2012	19.2	8.1
2015	11.8	5.3

Source: Michigan DTMB (www.milmi.org)

M3 Detroit – unemployment rate

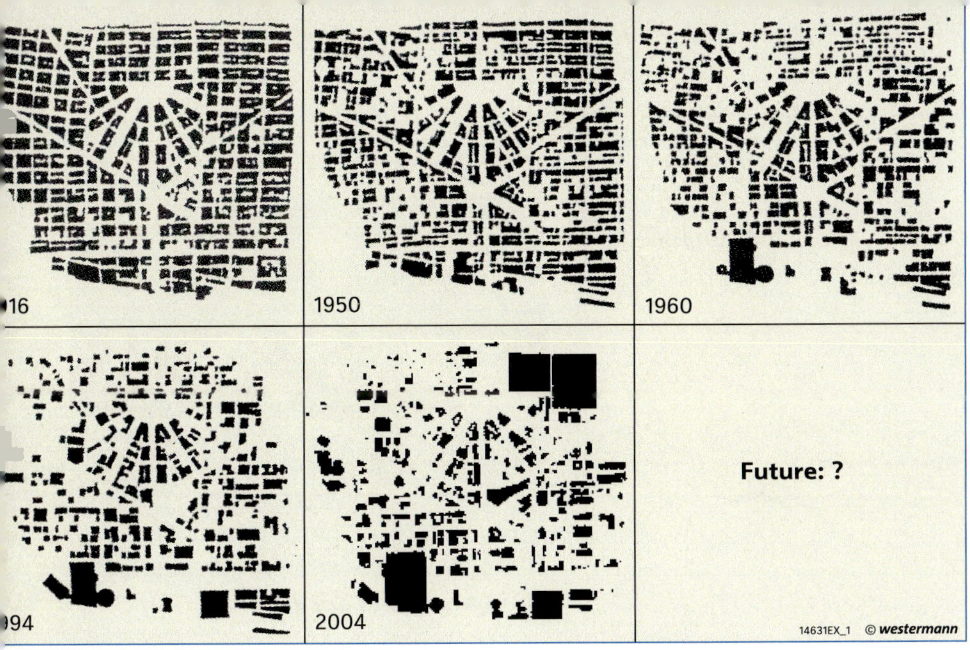

M4 Figure ground plans – changing downtown Detroit

M5 Struggling Detroit

100800-217-03
schueler.diercke.de

M6 Beach area in Campus Martius

Since the mid-1960s	Construction of shopping malls, offices and business parks along the main roads.
1977	Opening of the Renaissance Center, a skyscraper complex and headquarters of General Motors with hotels, shops, restaurants and offices.
Since 1999	Construction of three casinos in downtown with 38,000 daily visitors, offering more than 8,000 jobs.
2000 / 2002	Opening of Comerica Park Baseball Stadium for 42,000 spectators (2000) and opening of Ford Field Football Stadium with 65,000 seats (2002).
2004	Opening of Campus Martius, an urban center that offers offices, green areas and facilities for seasonal leisure amenities (ice rink, concerts).
2012	Quicken Loans, a mortage lending company, opened its headquarters in downtown Detroit. About 11,000 people work for the company.
2018	Microsoft will open a regional technology center in the Campus Martius building.

M7 Selected spatial changes in Detroit

TASKS

1. Describe the changes Detroit has undergone economically.
2. Explain reasons for and consequences of the industrial decline.
3. Evaluate the changes of Detroit's CBD.

KEY TERMS

- CBD
- downtown
- economic crisis
- Fordism
- revitalisation
- segregation

M1 Area of Lake Victoria

M3 Fishing on Lake Victoria

Lake Victoria – Once the Rich Fishing Grounds

100800-160 schueler.diercke.de

DE-160 www.diercke.com

pp. 32 - 33 Texts

Covering an area of almost 70,000 km^2, Lake Victoria is the second largest body of fresh water in the world, and the largest African lake. It fills a flat basin with a maximum depth of about 80 metres. An enormous number of fish species made the clear and light-flooded lake water their home. For a long time, the native fish provided food and a moderate but reliable income for the residents on the shoreline and the surrounding area.

Unleashing a 'beast'

Fishing on the lake began to intensify at the beginning of the 20th century. The British colonial power, needing many workers for its plantations, tried to boost the catch sizes for a fast growing population. This resulted in overfishing after a couple of decades, so the British officials decided to stock the lake with new, non-native fish. At the end of the 1950s, the Nile Perch was introduced to Lake Victoria. But as a strong predator, this fish caused the near or complete extinction of several hundred native fish species. These species had fed on the plankton, so the plankton had spread and enriched the lake with nutrients. This led to a decline in water quality after 1980. Ecologically, the introduction of the Nile Perch has changed the balance of Lake Victoria. Economically, a significant amount of this valuable fish is a potential resource.

Country	2008	2009	2010	2011	2012	2013
Kenya	4.0	3.1	3.6	2.8	3.1	2.9
Tanzania	23.3	17.9	16.3	15.0	15.3	12.4
Uganda	15.0	12.5	11.8	9.9	11.3	10.8
Total	**42.3**	**33.4**	**31.6**	**27.7**	**29.7**	**26.1**

Source: Eurostat (2013)

M2 EU imports of Nile Perch fillets (in 1,000 tonnes)

Boom and struggle in Tanzania

Life has become busy in Mwanza, Tanzania. Since the late 1990s, Tanzania has been exporting the Nile Perch worldwide, especially to Europe. Global overfishing has created a new demand for high-quality and inexpensive fish, especially in the rich countries. In this case, the Nile Perch has met these requirements.

Since the revenues from fishing Nile perch are significantly greater than those for traditional species, lots a people wanted to become a fisherman on Lake Victoria (**M5**). However, business is dominated by foreign fish companies. They offer jobs, but the work in their factories is moderately paid. Fishermen also complain about unfair trade practices with their catch.

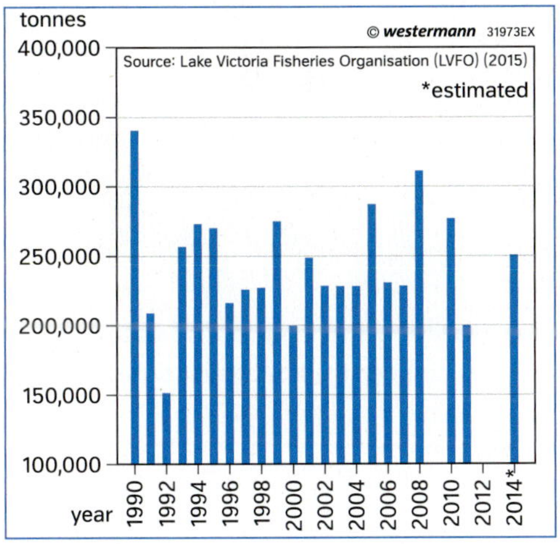

M4 Annual catch of Nile Perch

M6 Catch of juvenile Nile Perch

Large profits are usually made by the companies and their customers. Mwanza serves as one of the export centres. The frozen fillet leaves the country by plane. In Germany for example, the Nile Perch is known as 'Viktoriabarsch'. The fish remains stay in Africa and provide a small income as fish meal.

An uncertain future

As a big export product, the Nile Perch has become too expensive for many Tanzanians today because of rising prices. With the native fish species extinct or reduced by the perch, people here face an uncertain future. Meanwhile, the catch sizes of the Nile perch are also decreasing. The average weight per Nile perch caught fell from 50 kg in the 1980s to less than 10 kg today. Fishery was expanded to stabilize the annual catch. But overfishing is threatening the fish stock, and with it, one of the country's economic backbones. Furthermore, regional population growth is ecologically damaging due to increasing waste, sewage, deforestation, erosion, and water pollution. Colonialism and globalisation have ecologically degraded Lake Victoria. Unreliable as a food source, its economic potential is on the decline. However, positive aspects resulting from the fish exploitation such as an improved infrastructure can also be noted.

The government of Tanzania is struggling to establish a sustainable Nile Perch fishery on Lake Victoria. For example, the Specific Nile Perch Fishery Management Plan (NPFMP) tries to regulate and preserve fishery at the lake.

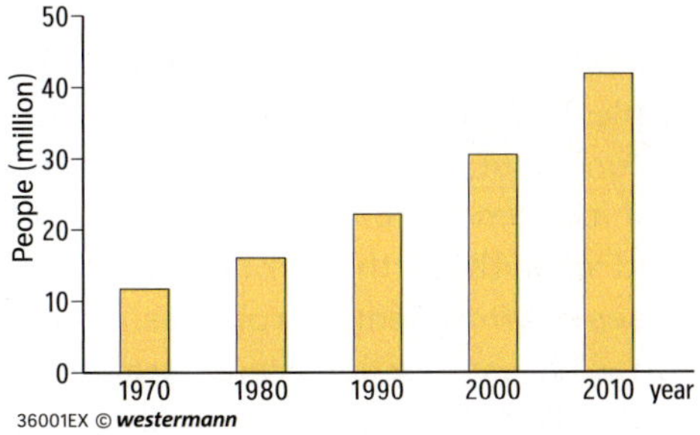

M5 Population in the Lake Victoria Basin

HELPFUL WORDS AND PHRASES
for TASK 3:

- The introduction of the Nile Perch had a positive effect on ..., but a negative effect on ..., because ...
- ... has a positive / negative environmental / social / economical impact ...
- I personally think ...

KEY TERMS

- deforestation
- erosion
- fishery
- overfishing
- pollution

TASKS

1. Make a timeline that shows the changes in fishery on Lake Victoria.
2. a) Explain the impacts the introduction of the Nile Perch had on the environment and on the people.
   b) For experts: Discuss the pros and cons of the introduction of the Nile Perch.

M1 District of Pudong (background) in 1990

M3 Pudong (financial centre Lujiazui) today

Shanghai – A Fast Changing City

100800-189
schueler.diercke.de

DE-108
www.diercke.com

Shanghai underwent a huge economic development in the past two decades. This progress was made possible by governmental reforms that established free-market economy principles and attracted international companies. One of the most important advancements was the creation of a special economic zone (SEZ) in Shanghai's Pudong district in 1990. Before then, the area of Pudong was used for agriculture. The SEZ propelled Shanghai to become China's economic and financial center. In 2013, a free trade zone was added to this special economic zone.

Special economic zone (SEZ)

An area with particular economic advantages, for example lower taxes compared to the rest of the country, to encourage investment and development there.

Free trade zone

A special area within a country where foreign companies can import materials, manufacture goods, export products, etc. without being limited by the usual rules and taxes.

FACT SHEET • Pudong (Shanghai)

- **Function:** Trade-, financial- and service center
- **Inhabitants:** >5 million (20 % of Shanghai's inhabitants)
- **Importance for Shanghai:**
  - Destination for one-third of all incoming foreign and domestic investments
  - 50 % share of Shanghai's foreign trade
  - 40 % share of Shanghai's GDP: 113.5 billion US-$
- **GDP per capita by comparison:**
  - Pudong: 22,000 US-$
  - Shanghai: 12,800 US-$
  - China: 8,069 US-$

M2 Basic information about Pudong (2015)

After the reforms, foreign investors were attracted because of favourable location factors like low taxes, free-trade, cheap land plots and a cheap labour force. In particular, logistics and service companies, as well as joint-venture industries producing for the global market took advantage of opportunities in Pudong. (**M4**).

Additionally, on an area of 520 km^2, new infrastructure was built, including a trade centre, the Zhangjiang High-Tech Park, universities, research facilities, a congress centre, leisure facilities and residential areas.

From 1990 to 2015, the GDP of Pudong increased from 873 million US-$ to 113 billion US-$, making Shanghai one of the richest cities in China.

Because of its economic importance, many people from the rural areas as well as from foreign countries are attracted by the city to which some also migrate illegally. This rapid growth of the urban population causes environmental problems such as high waste production, dense traffic and air pollution.

Joint venture

SAIC Volkswagen is an example of a joint venture between the foreign company Volkswagen Group and the Chinese company SAIC Motor headquartered in Anting, Shanghai. The Chinese partner offers the location, facilities and labor force. The foreign company provides investments and know-how. Both can profit.

Landmark

Shanghai's new skyscraper, the Shanghai Tower, located in the financial district of Pudong, is the second largest building in the world (632 metres). It was opened in 2016. It offers offices for multinational companies and financial service providers, luxury hotels, high-end stores, as well as leisure- and event facilities.

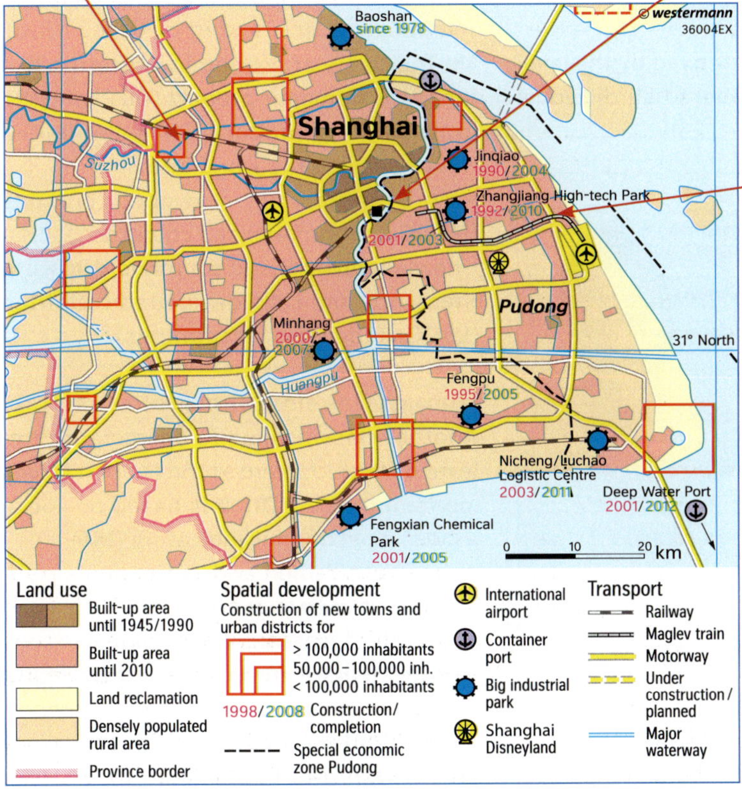

M4 Development of Shanghai and chosen locations

pp. 10-11 *Thematic Maps*

Infrastructure

With the world's largest container port, two airports, a magnetic levitation train that connects Shanghai International Airport with the free trade zone Pudong, and various other train and road networks within the city, Shanghai offers very good conditions for foreign trade.

TASKS

1. Describe Shanghai's development since the 1990s in five statements.
2. Explain why Shanghai is attractive for foreign investors.
3. Comment on the importance of Pudong for Shanghai.

KEY TERMS

- foreign direct investment
- free-market economy
- free trade zone
- joint venture
- special economic zone

HELPFUL WORDS AND PHRASES

for TASK 3:

- Due to governmental reforms ...
- Favourable location factors for foreign investors are ...

M1 A classroom in Tanzania

M3 A classroom in England

Level of Development and Indicators

100800-270, -274 schueler.diercke.de

DE-192 www.diercke.com

Different countries of the world can be characterized by the level of development. In order to do so, geographers use different indicators.

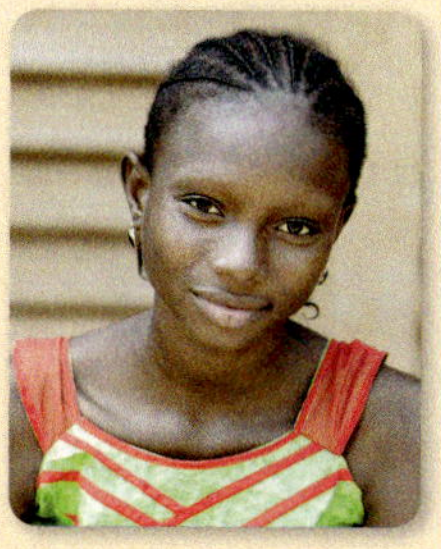

'Hello, my name is Kibwana. That's Swahili, the language most people speak in Tanzania. I live in a small village at the foot of the Kilimanjaro with my mum and five brothers and sisters. My father died when I was little. Next to our small house we plant a few crops to have a little bit to eat, and we are very proud of our goat. I have to work in the fields every day to support my family and help to get some food and water, so I can't go to school. I've heard that they want to build a new school in another village not far from here, and I'm hoping to have the chance to go there one day.'

***

'Hi, I am Meagan, and I live in Cambridge, England with my mother and father. We moved here into a new, big house because my father got a new job as an IT consultant. I like it here, and I also found new friends already. We like to go to the movies and concerts a lot. After I am graduated from school, I want to study medicine at a good university. But before, I really would love to travel around the world. There are so many nice places to see! But for now, I'm looking forward to the next summer holidays where we will travel to Los Angeles for 3 weeks.'

M2 Two children tell about their living conditions

Measuring economic development

On the basis of economic indicators like the gross domestic product (GDP) and the gross national product (GNP), the United Nations divide the world into:

- more economically developed countries (MEDC, or industrialised countries),
- newly industrialised countries (NIC) and
- less economically developed countries (LEDC, or developing countries).

Human Development Index

Another classification to characterise the level of development is the Human Development Index (HDI). It is based on social as well as economic indicators (**M4**):

- health, nutrition and hygiene: *life expectancy at birth*,
- standard of living: *GNP / capita*,
- education: *average school attendance in years and expected length of school attendance in years*.

The UN has been publishing the Human Development Report (HDR) including the HDI annually since 1990.

Gross national product (GNP)
The total value of all the goods and services produced by a country in one year, including the total income from foreign countries.

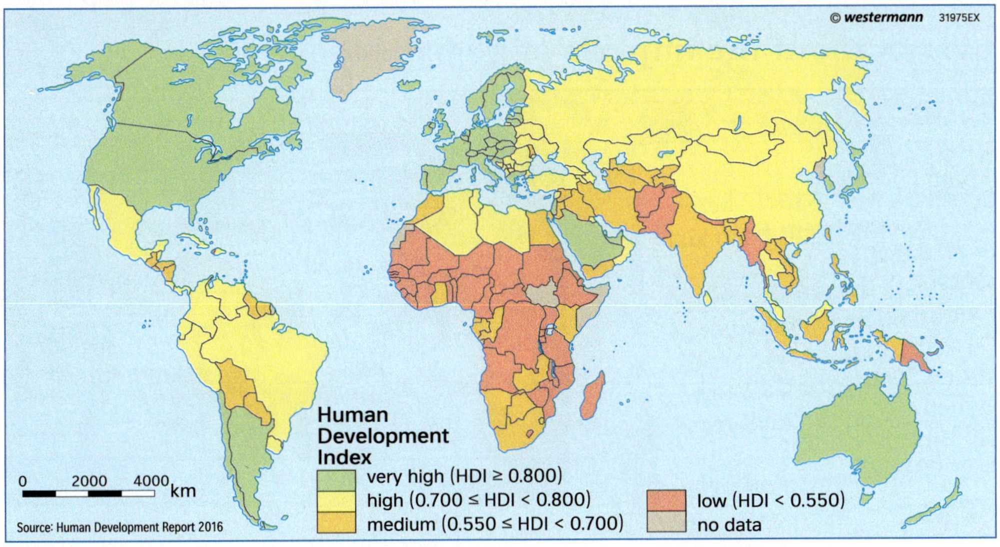

M4 Level of development of countries according to the HDI (2015)

pp. 16-17 *Tables*

HELPFUL WORDS AND PHRASES
for TASK 2:

- Most developed / developing countries are located in ... ... can be located / found ...
- Except Australia, most developed / developing countries are located north / south of the Equator.

Indicator	Tanzania	China	United Kingdom
Inhabitants (million)	51	1,367	64
GDP (billion US-$)	46.19	11,380	2,865
GNP per capita (current international US-$)	2,630	14,390	40,900
GDP, by sector of origin (2014, in %): agriculture/industry/services	26.5 / 25.6 / 47.3	8.9 / 42.7 / 48.4	0.6 / 15.2 / 83.5
Exports (billion US-$) most important goods	5.4 gold, coffee, cashew nuts, manufactures	2,270 electrical & other machinery, computers	442 manufactured goods, fuel, chemicals, food
Imports (billion US-$) most important goods	10.5 consumer goods, machinery & transportation equipment, industrial raw materials	1,596 electrical & other machinery, oil & mineral fuels, nuclear reactor, machinery components	617 manufactured goods, machinery, fuel, foodstuffs
Life expectancy (years)	61.7	75.4	80.5
Literacy rate (age 15 and over in %)	70.6	96.4	99.0
Population below poverty line (%)	67.9 (2011)	6.1	1.5 (2013)
HDI (rank of 188)	151	90	16
Gender development index	0.937	0.954	0.964

Source: CIA Factbook, World Bank (2015)

M5 Structural data of different countries

TASKS

1. Contrast the photos (**M1**, **M2**).
2. Describe the distribution of countries with a high level of development and a very low level of development (**M4**).
3. Make a list of the characteristics of more developed countries and less developed countries using **M3-M5**.
4. Discuss the importance of some of the indicators used in **M5** for the level of development. Answer in English or German.

KEY TERMS

- development
- gross national product (GNP)
- HDI
- LEDC
- life expectancy
- MEDC
- NIC
- standard of living

Different views about the impacts of globalisation:

'In Globalization 1.0, which began around 1492, the world went from size large to size medium.
In Globalization 2.0, the era that introduced us to multinational companies, it went from size medium to size small.
And then around 2000 came Globalization 3.0, in which the world went from being small to tiny.'

Thomas Friedman (US American journalist and author)

'Where globalization means, as it so often does, that the rich and powerful now have new means to further enrich and empower themselves at the cost of the poorer and weaker, we have a responsibility to protest in the name of universal freedom.'

Nelson Mandela (South African activist against Apartheid and former President; 1994 - 1999)

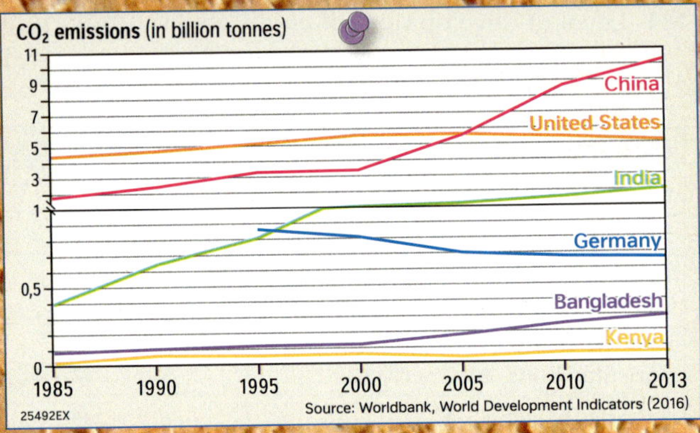

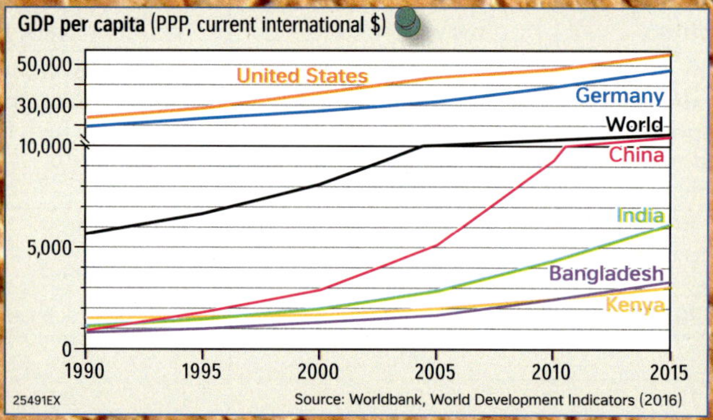

'Globalization, as defined by rich people like us, is a very nice thing ... you are talking about the Internet, you are talking about cell phones, you are talking about computers.
This doesn't affect two-thirds of the people of the world.'

Jimmy Carter (President of the USA; 1977 - 1981)

M1 Different views about the impacts of globalisation

Globalisation – Different Views

pp. 18 - 19 Line Graphs
pp. 32 - 33 Texts
pp. 154 - 155 Cartoons

Globalisation affects politics, economics, culture, societies and the environment. The consequences, however, have led to controversial discussions. Supporters argue that globalisation offers chances for all countries to develop economically and reduce unemployment and poverty. Opponents are afraid that only the rich profit from a globalised world, while the poor become even poorer. They also fear a negative impact on local enterprises and cultures, as well as on the environment.

Positive effects of globalisation:
- Free trade reduces (tariff) barriers
- Global economic growth (imports and exports)
- Possibility of participation in global world trade, also for companies in LEDCs
- Creation of jobs
- Lower prices for consumers
- Availability of products from different countries

Notes from a Geography lesson

The trends that are shaping the world today embody both chances and risks. Globalisation has lifted millions of people out of poverty while contributing to social fragmentation, a massive increase in inequality, and serious environmental damage.

Notes from an interview with a economist

Labor force - negative impacts of a globalised world

MEDCs:
- Transfer of production to lower cost countries → job losses

LEDCs:
- Unfair/bad working conditions
- Exploitation of labor force
- No social security
- Low wages

Notes from a Geography lesson

'If I had to capture the meaning of globalization in one word, I would say 'uncertainty' – and that's something that will increase rather than decrease in the future.'

Ivan Miklos (former Deputy Prime Minister of Slovakia)

TASKS

1. Make a list of the pros and cons of globalisation. Use the text and the materials on the pinboard (**M1**).
2. Comment on one of the opinions.
3. Analyse the cartoon.
4. Discuss: Globalisation – blessing or curse? Answer in English or German.

KEY TERMS

- environment
- fragmentation
- globalisation
- multinational company
- poverty

HELPFUL WORDS AND PHRASES

for TASK 4:
- Weighing up the pros and cons ...
- According to the different views ...
- With reference to the various arguments ...

Pedestrians

People

42 World Population
44 The Demographic Transition Model
46 The Challenges of Population Change
48 Feeding the World
50 International Migration
52 Internal Migration

M 1 Multicultural New York

M1 *'Don't you think, it's gettin' a bit too tight here?'*

World Population

Never before have so many people lived on Earth. About 7.45 billion people inhabited the planet at the end of 2016. Every minute the world population grows by 171 people.

Growth of world population

The fast growth of the world population is a recent phenomenon in the history of the world (**M2**). For a long time populations did not grow significantly. Although birth rates were high, the natural increase remained modest because death rates were also high. The rate of growth picked up in Europe in the 19th century following health and hygiene improvements in the course of the Industrial Revolution. The very rapid growth of the world population started with a sharp reduction in death rates in the less economically or even least developed countries (LEDCs, LDCs, **M3**).

However, since the 1960s there has been a change. Global population growth has slowed down due to a decline in fertility in most parts of the world. Today, the world population is increasing at a rate of 1.2 per cent per year. This rate is expected to decrease even more to about 0.5 per cent by 2050. Despite this downward trend, the world population will continue to grow in the 21st century, but at a slower rate.

Population distribution

The population is distributed very unevenly across the world. Natural conditions are one reason. Hostile environments such as hot and cold deserts or mountain areas are sparsely populated. In contrast, areas with fertile soils and favourable climatic conditions may show large population numbers.

Birth rate
is the number of live births per 1,000 in a population per year.

Death rate
is the number of deaths per 1,000 in a population per year.

(General) Fertility rate
is the number of live births per 1,000 women aged 15 to 45 per year.

Rate of natural increase
is the difference between the birth rate and the death rate of a country or place. Due to this rate there is a natural increase or decrease.

Migration
is the movement of people in and out of an area, for example to find work and/or better living conditions.

Population growth (rate)
is the development of the number of people due to natural increase or decrease and migration expressed as a percentage of the population.

	Population (millions)	Births per 1,000 population	Deaths per 1,000 population	Rate of natural increase	General fertility rate
World	7,336	20	8	1.2	2.5
MEDCs	1,254	11	10	0.1	1.7
LEDCs*	6,082	22	7	1.4	2.6
LDCs	938	34	9	2.5	4.3

Source: www.weltbevölkerung.de (2015), *including LDCs

M3 Population growth worldwide and by groups of countries

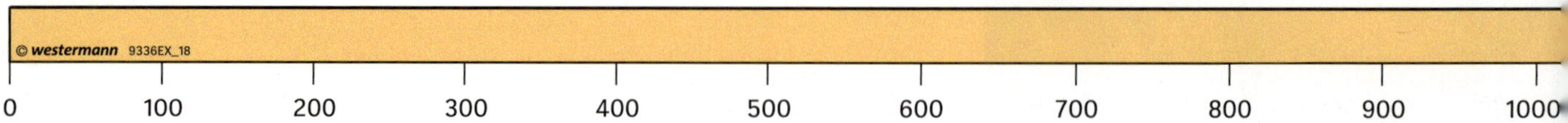

M2 Global population growth

Another reason for the differences in population distribution are the varying rates of population growth across the world (**M3**, **M4**). LEDCs and LDCs still have high birth rates. In these countries, women traditionally marry early and begin bearing children at a much earlier age than in more economically developed countries (MEDCs). As a result, the share of people living in economically less developed countries has constantly risen from 65 per cent in 1970 to 83 per cent in 2015.
Apart from different natural increase rates, regional differences in distribution also depend on migration (› pp. 50-51).

www.worldpopulationhistory.org

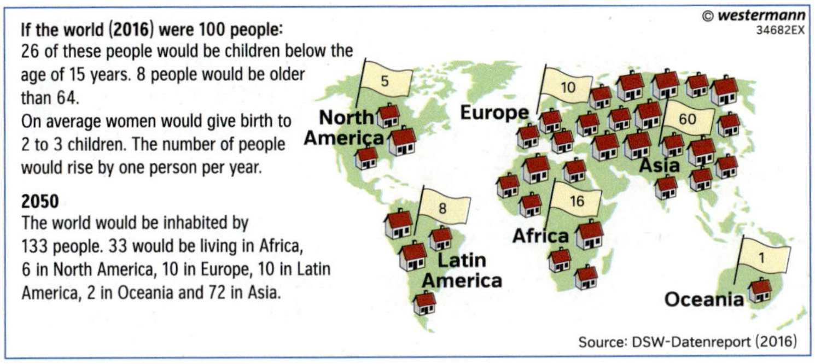

M4 If the world were 100 people, ...

100800-276 schueler.diercke.de

DE-190 www.diercke.com

TASKS

1. a) Describe the development of the world population until 2016 and the projections for 2050 (**M2**).
   b) Discuss possible reasons for the different fertility rates per woman as projected in **M2**.
2. Compare the population development in MEDCs, LEDCs and LDCs (**M3**).
3. a) Make a diagram which shows the changes in the regional population distribution from 2017 to 2050 (**M4**).
   b) Calculate the population growth for the different regions in **M4** (as a percentage of 2016).
4. Give reasons for the differences in population distribution (text, atlas).

KEY TERMS

- birth rate
- death rate
- general fertility rate
- LEDCs / LDCs / MEDCs
- population distribution
- population growth
- population growth rate
- rate of natural increase

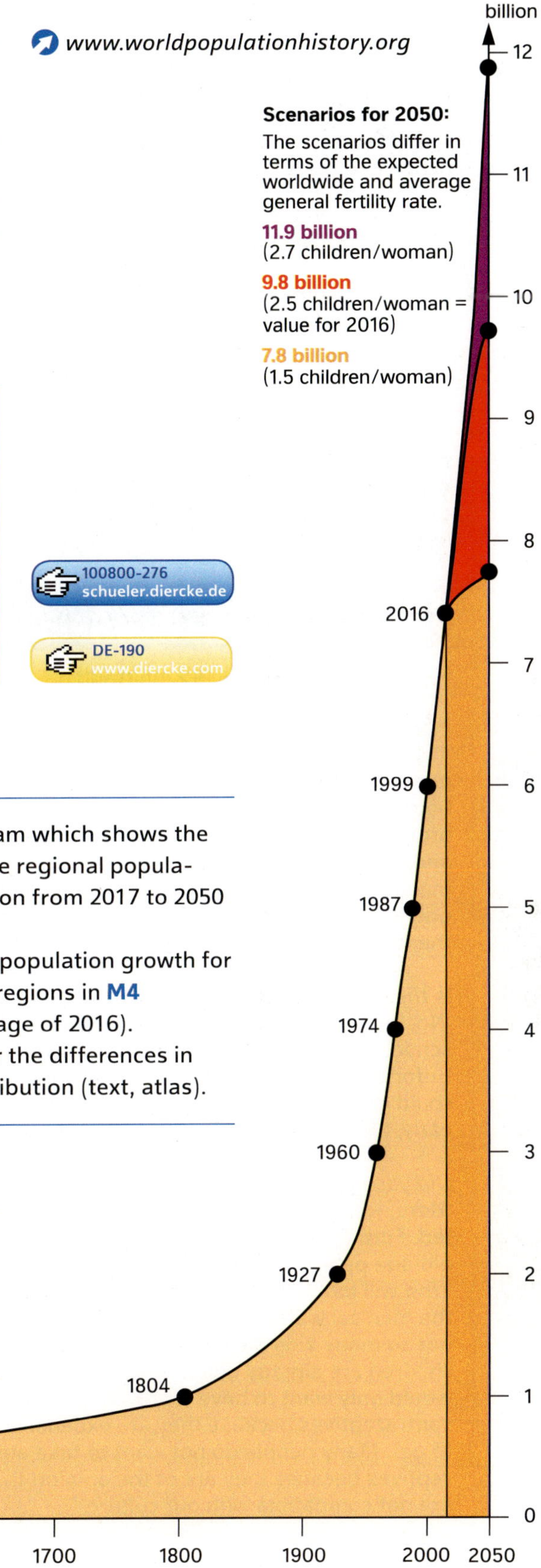

The Demographic Transition Model

100800-276 schueler.diercke.de

DE-190-1 www.diercke.com

pp. 152 - 153 Age-and-Gender

The Demographic Transition Model (DTM) shows how birth and death rates change within a country. It reflects changes in the economy and in living conditions.
The model was originally used to describe historical population developments in Europe and North America in the past 150 – 200 years.

Despite its limitations, the DTM is being used to portray and explain current population developments worldwide.

Madiha from Tanzania (45, married)

Please tell us something about your life situation.
'I've got six children and two grandchildren. That's not unusual. My sister has got seven and my neighbour eight kids. We've got a piece of land where we grow cereal. In addition, we've got chickens, cows and goats. Our farm is not very profitable, which means we cannot employ any workers. As a consequence, our children have to work with us to support the family.'

Is there a school here in the village?
'No, but in the neighbouring village we've got a school, which is about two hours away on foot. Unfortunately, the school fees are expensive, and we could only send the boys to school. The girls marry early, that's why school is not important for them.'

Did you always want to have so many children?
'Here, women with many children are well respected. Furthermore, in Tanzania we do not have any old age pension plans or provisions for the elderly. Who will look after me when I'm ill? Since we've had the medical ward and the children get vaccinated, not so many kids die any more. But still, two of my children are already dead. If I could choose today, I would only want to have four kids. Only few use contraceptives because they are expensive and hard to get. Many people do not want to take any contraceptives because they are of the opinion that they are not compatible with our culture.'

Ursula from Switzerland (35, married)

You have a daughter and a job. Is it easy for you to reconcile your job with your family?
'In the first year it was quite difficult, because I wanted to spend much time with my daughter and yet not jeopardize my career. Now we have a day care place, and my husband supports me.'

Did you always want to have kids?
'No, not at all. Deciding for a child was not easy for me. I thought about it for a long time. I was independent and had a good and well-paid job as a lawyer. I was able to afford a lot and to realise some of my dreams. Then I got to know my husband and was ready to start a family. I could choose when to have a child.'

Would you like to have a second child?
'No, I don't want to have a second child. Nowadays, you have more control due to contraceptives. My grandma used to have four kids. I myself have two siblings.'

Why do you think that's so?
'The children do not need to contribute to the family income. Life expectancy is much higher, but there's retirement money and there are pensions. Children cost a lot of money. I think that many couples worry that they won't be able to maintain their living standards. Having children restricts gainful employment, at least for one partner.'

M1 Two women from different countries tell about their family life

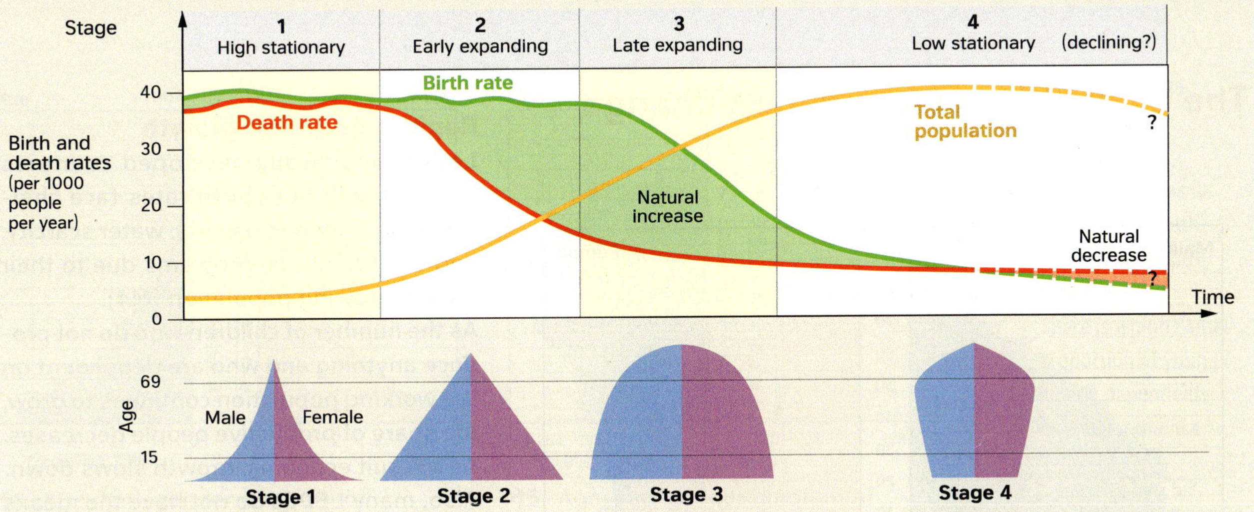

Factors influencing

	Stage 1	Stage 2	Stage 3	Stage 4
Birth rate	• religious beliefs • children needed for work in farming • high birth rates make up for high infant mortality rates • lack of contraceptives and family planning		• lower infant mortality • mechanisation reduces need for workers	• improved status of women (education) • family planning • later child birth • later and fewer marriages • wealth chosen over large families (children are expensive)
Death rate	• diseases • famine • poor sanitation • poor medical knowledge	• improved health care (e.g. smallpox vaccine) • advances in hygiene • improved sanitation • improved water and food supply		• medical advances (e.g. transplants) and preventative medicine • better food supply

M2 Demographic Transition Model (DTM), factors influencing birth and death rates, and typical shapes of age-and-gender diagrams according to stages in the DTM

Country	Births per 1,000 population	Deaths per 1,000 population	Population growth rate (%)
Germany	8.5	11.6	-0.2
Switzerland	10.5	8.2	0.7
China	12.4	7.7	0.4
India	19.3	7.3	1.2
Tanzania	36.0	7.8	2.9
Mali	44.4	12.6	3.0

Source: The World Factbook (2016)

M3 Demographic data of selected countries

TASKS

1. a) Compare the lives of Madiha and Ursula (**M1**).
   b) Give reasons for the differences in family size.
2. a) Describe the different stages of population development (**M2**).
   b) Explain the different stages of the Demographic Transition Model DTM (**M2**).
3. Assign four countries from **M3** to the appropriate stage in the Demographic Transition Model (DTM).

KEY TERMS

- birth rate
- death rate
- Demographic Transition Model
- infant mortality rate
- life expectancy
- population growth rate

HELPFUL WORDS AND PHRASES

for TASK 2:

- The first stage is characterised by a high / low / fluctuating / steady / falling birth / death rate.
- In the second / third stage ...
- The birth / death rate begins / continues to fall ... slowly / rapidly / below ...
- The birth / death rate changes from ... to ...
- As a result, the natural increase is becoming low / high
- The population growth is expanding / stationary / declining because ...

M1 Cartoon

'Everyone has the right to a standard of living adequate for the health and well-being of himself and his family, including food.'

Universal Declaration of Human Rights (U.N. 1948), Article 25

Food security exists 'when all people at all times have access to sufficient, safe, nutritious food to maintain a healthy and active life.'

World Food Summit (1996)

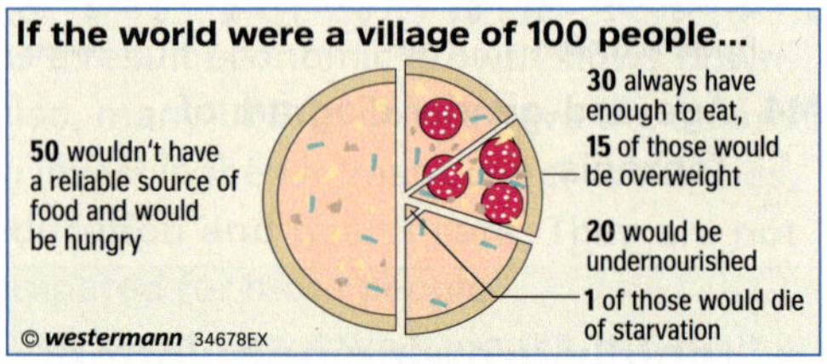

M3 How many have enough to eat?

Feeding the World

Malnutrition and hunger

People suffer from malnutrition or undernutrition if they are persistently hungry because they either do not take in enough calories over a long period of time (chronic hunger), or they do not get enough vitamins and minerals to live healthy lives (hidden hunger).

Malnutrition can also be the result of overnutrition. People who consume too many calories and / or have an imbalanced diet are malnourished, too.

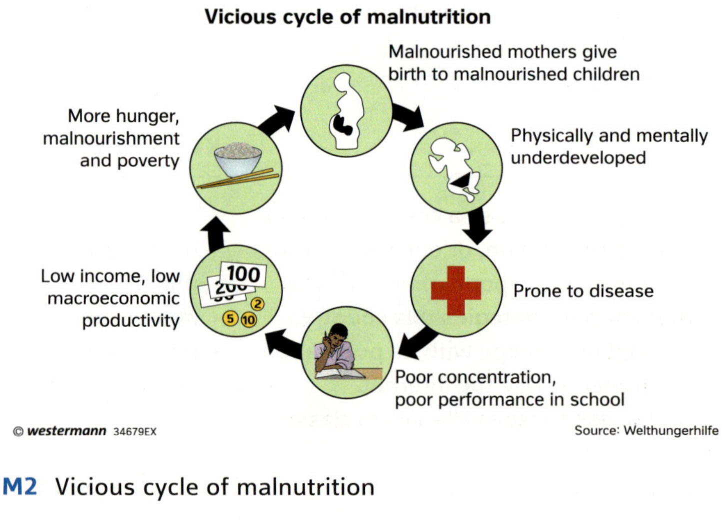

M2 Vicious cycle of malnutrition

Malnutrition – a global problem

One in seven people on Earth is suffering from chronic hunger despite the fact that we are producing about one and a half times the amount of food needed to provide everyone in the world with an adequate diet.

In addition, there are many people who are affected by hidden hunger. If we add to these two groups those people who are overnourished, roughly half the world's population does not receive an adequate diet (**M1**, **M3**).

The global food gap

Many of the overweight live in the rich parts of the world, i. e. in more economically developed countries (MEDCs) and in newly industrialized countries (NICs).

By contrast, the majority of people suffering from undernutrition live in less economically developed countries (LEDCs, › pp. 36-37). They are the poor who do not have the resources – whether land, tools or money – that are needed to grow or buy food.

Lack of healthy food and proper nutrition affects every stage of life, trapping people in a vicious circle that passes hunger from one generation to the next (**M2**).

Feeding the world today and tomorrow

Global crop production
Over the past 50 years, crop production has increased significantly due to improved cropping techniques and technology, use of fertilizers and pesticides, as well as irrigation.

Population growth
The world's population is projected to grow to 9 billion before 2050. More than half of this growth will occur in sub-Saharan Africa, a region where one-quarter of the population is undernourished.

Land use
According to the United Nations Development Programme, if land use continues to increase, the world will already have reached the limits of ecologically sustainable land use by 2020.

Urbanisation
By 2050, urban areas will have increased by about 70 per cent and will cover about five per cent of the world's land area. Arable land has to give way to towns and cities.
To this day, people in cities produce little to none of their own food, and particularly the urban poor often lack the means to buy all they need.

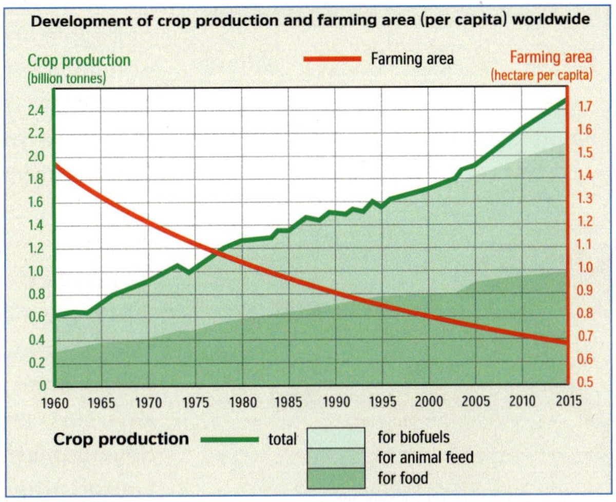

M4 Global crop production (1960 – 2015)

Climate change
As a result of global warming, the number of extreme climatic events such as floods and droughts is expected to increase.
This will negatively impact crop yields, particularly in those parts of the world where hunger is most prevalent.
However, modern agriculture also contributes to global greenhouse gas emissions, and accounts for 70 per cent of all freshwater withdrawn from rivers, lakes, and aquifers.

Mass production of meat and dairy products
Rising affluence and urbanisation in some emerging countries are bringing about a higher demand for a more western diet, i. e. for convenient, processed foods, for meat, and for dairy products. However, more people can be fed from a given area of arable land if it is not used to feed livestock or be an energy crop for biofuel production.

Biofuels
Renewable sources of energy such as biofuels are said to reduce climate change and dependency on non-renewable resources of energy.
They receive a lot of political support from governments in more economically developed countries (MEDCs).

100800-275 schueler.diercke.de

DE-192 www.diercke.com

pp. 18 - 19 *Line Graphs*

pp. 154 - 155 *Cartoons*

TASKS

1. Explain the phrase: 'Malnutrition – a global problem'. Use your atlas.
2. a) Describe the vicious cycle of malnutrition (**M2**).
   b) Explain why it is difficult to escape from it.
3. a) Describe developments shown in **M4**.
   b) Give reasons for these developments.
4. For experts:
   a) Collect further information on one of the factors above.
   b) Make a poster to illustrate and to present your information.

KEY TERMS

- food security
- hunger
- malnutrition
- overnutrition
- undernutrition

Distribution of international migrants by origin and destination (2015)

Direction			Stock (million)	Percentage of total international migrant stock
South	→	South	90.2	37 %
South	→	North	85.3	35 %
North	→	North	55.2	23 %
North	→	South	13.6	5 %

Source: UN DESA (2015)

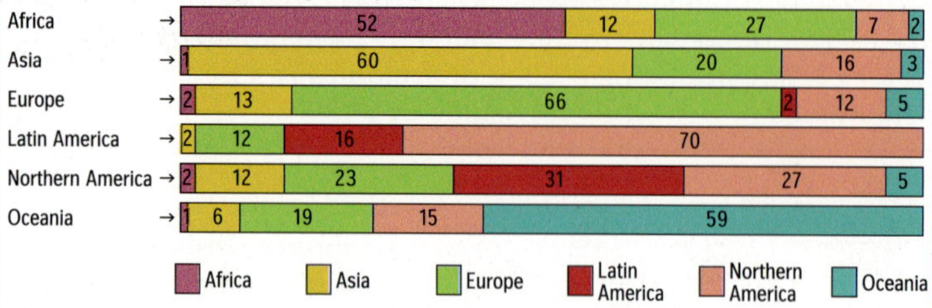

Source: United Nations

M1 Distribution of international migrants

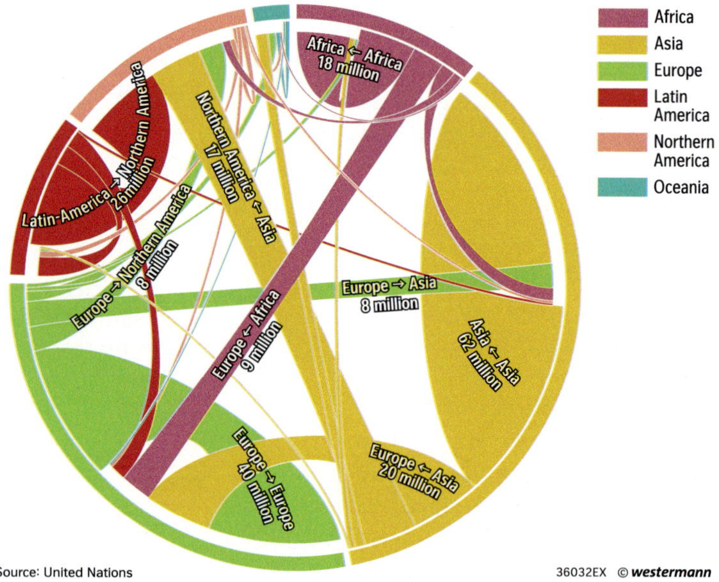

Source: United Nations

M2 Number of international migrants (2015)

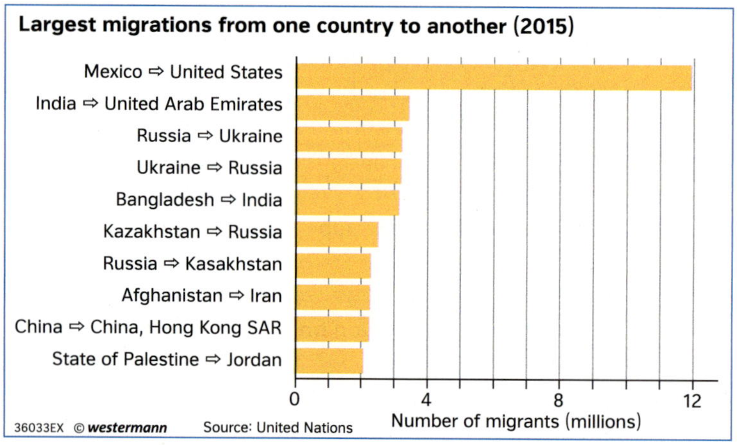

Source: United Nations

M3 Important populations of migrants

International Migration

Across the world, millions of people are on the move. Today, about 250 million people live outside their home countries, a number that has steadily increased.

People migrate for many different reasons. There are usually push and pull factors at work (**M4**, › p. 52). Most migrants hope to live a better life or to find a (better) job (labour migration). An increasing number of people are forced to migrate due to war and environmental problems.

Therefore, major international migration flows are from developing countries to other countries that are better off (**M1**-**M3**).

Impacts on the home country

Often the motivated, young and educated people move away. This deprives their home country of much-needed skilled workers such as doctors and engineers. This development is called brain drain.

On the other hand, many families benefit from money sent home by migrants. This money constitutes an important source of income for the families and for their country as well, because it is spent on consumption, education and health care.

In addition, many migrants return back home after a few years abroad. The return of highly trained migrants with specialized knowledge and skills can trigger economic and social developments in their home country.

Migration

Migration is the movement of people from one place to another. It can be voluntary or forced, legal or illegal, internal (within a country) or international (crossing state borders), permanently or temporarily. People leaving their home country to move to another place are called emigrants. In the country they move to they are called immigrants.

M4 Push factors and pull factors

Impacts on the host country

Many more economically developd countries (MEDCs) have restrictions on immigration, fearing rising competition for jobs and housing, an increased demand for social services, integration problems, as well as social and intercultural conflicts.
On the other hand, there is often a demand for unskilled workers in the hospitality, building and farming sectors where immigrants frequently work illegally for low wages.
Host countries also benefit from highly-skilled migrants, which is called brain gain. Many MEDCs are competing for educated and trained workers since job vacancies (e.g. in the IT sector) cannot be filled by the home population alone.

What is more, many western countries have to cope with an ageing population (› pp. 46-47). Thus, the influx of young migrants positively impacts their population structure and helps stabilise their social systems.

100800-279
schueler.diercke.de

DE-195
www.diercke.com

pp. 20 - 21
Bar Charts

Push factors and pull factors

Push factors are reasons that make people leave a region or country.
Pull factors, on the other hand, attract people to a new place. These factors can be economic, social, political or environmental.

TASKS

1. Name important target regions and regions of origin of migration (**M1-M3**).
2. a) Prepare a presentation about particular migration flows and their background (**M4**).
   b) Summarize your findings.
3. For experts: Discuss the impacts of labour migration on both home and host countries. Answer in English or German.

KEY TERMS

- brain drain
- brain gain
- emigration
- immigration
- migration
- pull factor
- push factor

HELPFUL WORDS AND PHRASES
for TASK 3:
- ... has postive / negative impacts on ...
- On the one hand, ..., on the other hand, ...
- ... which means (that) ...
- As a result, ...
- However, ...
- Thus ...

Internal Migration

100800-279 schueler.diercke.de

DE-195 www.diercke.com

Internal migration is the movement of people from one place to another within a particular country. People move from cities to suburbs and vice versa, or between cities and rural areas. Their reasons for moving are manifold (**M1**-**M3**).

Trends

In less economically developed countries (LEDCs), poverty and poor living conditions in rural areas are important push factors for people to move into cities. They believe that there is the chance for a better life there. Rural-to-urban migration is a major driving force of city growth. Consequently, the world is steadily becoming more urban (**M4**). Already, more than half the world's population lives in cities. In 2050, it is estimated that two-thirds of the world's population will live in urban areas.

'I had lived in Jersey City my whole life before moving to this small suburban town. But Jersey City has gotten so dirty. Besides, it is also becoming too crowded. My husband had always wanted to get out of the city and have wide-open space. In Jersey City after coming home from work, he used to drive around the neighborhood for at least an hour looking for a place to park. Now we have our own driveway. And I have my own nice garden. I am tired of the hustle and bustle in the city. I just want some quiet. And here's not as much crime as in the city. We would never have been able to afford a place like this in Jersey City.'

Angela Smith (61 years, USA)

M1 Migrant's report from the USA

'Before we moved from the suburb to the city of Manchester, I had to commute for some time, which meant that my kids had already gone to bed when I came home. I only saw them on the weekends. Now the commute is much shorter, and I can see them before going to work, take them to school, and be with them for an hour before they go to bed. I love the idea of raising kids here, because it is such an exciting place! There's so much to do! In the suburb where we used to live, I sometimes felt isolated and bored. The city provides the whole family with easy access to jobs, better contacts, museums, schools, and so on. It's all at your doorstep!'

Malcom Donovan (35 years, UK)

M3 Migrant's report from the UK

'I left my village because there are no employment opportunities there, which is what made me come to Nairobi. I could not live well back home, could not help my mother, so I came here in search of a better life. I came here without having a place to stay, but I knew of someone from my village who I could stay with. My first impression of Nairobi was that this was a place with better opportunities. There are lots of nice things: there's electricity, whereas we have none in our village, transportation, and jobs are easier to find. I work as a housekeeper. Each month, I send about $20 of my monthly salary of $115 back home, about 200 kilometres away. However, I'm sad that I can only see my family once a year when I'm on vacation. My long-term goal is to go to school and learn a skill which will allow me to earn a better salary. Then I can help my mother the way I'd like to.'

Sarah Mwende (29 years, Kenya)

M2 Migrant's report from Kenya

People

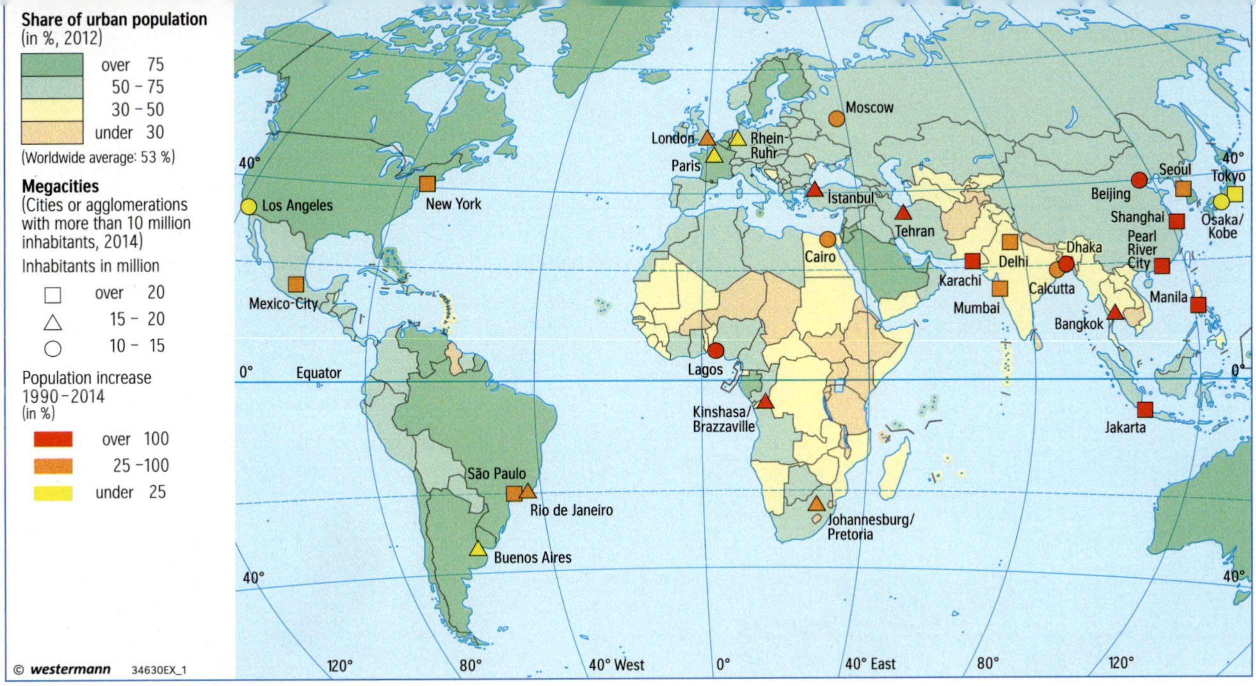

M4 Share of urban population and megacities

The trend in city growth is towards ever-larger agglomerations. Today, one of ten people lives in one of these megacities with more than 10 million inhabitants.
However, this trend is not uniform. In LEDCs, where most of the population still lives in rural areas, growth in the urban centres has been faster than in MEDCs, where a lot of people already live in cities. In fact, some of them want to leave the city in order to live in the surrounding area. This is called suburbanisation (**M1**).

Problems of growing cities

Many cities in LEDCs cannot keep up with the increasing number of people, who often cannot find jobs to provide for their families.
As a consequence, urban poverty and homelessness have become serious problems. Almost one billion urban dwellers live in informal settlements. The housing is of such poor quality and with such inadequate provision of water, sanitation, drainage and garbage removal that their health is under constant threat.

Urbanisation

The term urbanisation is used differently in German and English. The English term covers two aspects:

1. The growth of towns in size, an increase in the number of towns and in the number of people living in towns. In German geography, this phenomenon is referred to as *Verstädterung*.
2. The spread of urban styles of living into rural areas. The German geographical term for this is *Urbanisierung*.

TASKS

1. List the push and pull factors mentioned in the reports (**M1-M3**).
2. a) Name regions with a high share of urban population (**M4**).
   b) Point out regions where cities grew fast from 1990 to 2014 (**M4**).
3. Discuss the causes and problems of urban migration (**M2**, text).

pp. 10-11 Thematic Maps

HELPFUL WORDS AND PHRASES
for TASK 3:

- There are various reasons why ...
- On the one hand ..., on the other hand ...
- Many people are driven by ...
- Due to ...
  ... has an impact on ...
  ... affects...
  ..., which means that ...
- As a result, ...
- While ...

KEY TERMS

- agglomeration
- megacity
- migration
- rural-to-urban migration
- suburbanisation
- urbanisation

Clouds
Smoke
Chimney stacks
Cooling towers
Solar farm

Power cords
Pylon
Solar panels
Penstock

Energy Resources

56 Energy – A Vital Component of Our Life
58 Coal – The Dirty but Indispensable Fuel
60 Crude Oil – Still the Black Gold
62 Natural Gas – A Clean Fossil Fuel?
64 Nuclear Energy – Friend or Foe?
66 Hydropower – Electricity from Water
68 Bioenergy – Not Just Firewood
70 Energy – Blown by the Wind
72 Solar Energy – Let the Sunshine In
74 Other Renewable Energies
76 World Energy – Challenges and Perspectives

M 1 Energy mix in Germany

Energy – A Vital Component of Our Life

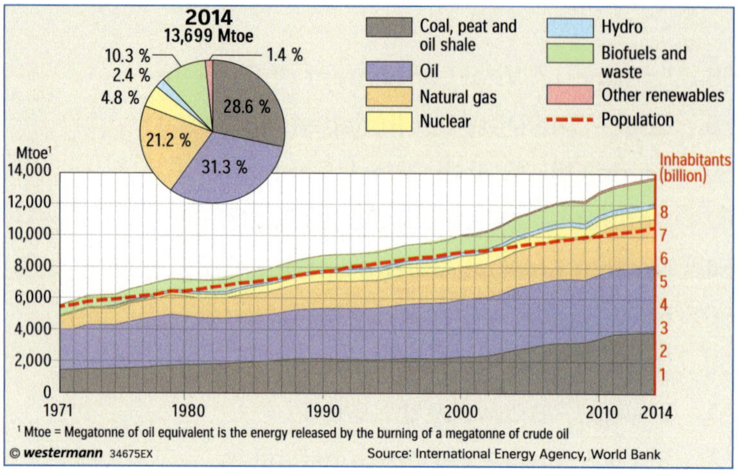

M1 World energy production and population

Much of our well-being depends on the availability of energy. We need this energy not only to drive machines, but also to heat and generate electricity. All the new things that make our lives easier and more enjoyable need energy – not least to manufacture them. Economic growth and the rapidly growing population of the earth leads to an increasing demand for, and use of, energy resources (**M1**). Recovery, production and use of energy are often connected with long-lasting effects on the environment.

Non-renewable energies

Most of our energy comes directly or indirectly from three sources: Coal, crude oil and natural gas owe their existence to plants and animals that lived millions of years ago. These became fossilised and are imbedded in rocks far beneath the earth's surface. They still contain much of the energy from the biomass they were made of. Therefore, they are called fossil fuels.

Their energy cannot be replaced once we burn these non-renewable energy resources. Coal, crude oil and gas deposits become more and more depleted and the resources will eventually run out. When this will be the case is very much disputed.

Moreover there are environmental problems. When the energy of fossil fuels is released, harmful gases and particulates pollute the atmosphere and contribute to climate changes (› pp. 80-81). Oil and natural gas production, coal mining and the transport of fossil fuels to the point of use can have environmental impacts as well. Another non-renewable energy resource is uranium which fuels nuclear power stations. In the production of electricity from uranium, no greenhouse gases are released. There are, however, extremely harmful long-term radioactive waste products which are difficult to store.

Renewable energies

Most of the renewable energy resources rely on the sun's energy. This is most apparent in solar panels, where the sun's energy is converted into heat or directly into electricity. For hydropower, the sun drives the water cycle so that precipitation coming as water from the mountains through rivers drive turbines to produce

Resources and reserves

The primary fuels – coal, crude oil, natural gas, and uranium – are material resources. A resource is the total amount of a material that exists. This includes fuel that is both discovered and undiscovered, economically recoverable or not economically recoverable.

On the other hand, reserves are deposits of fuels that are known to exist based on geological and engineering studies. Reserves can be recovered economically with the technologies that already exist.

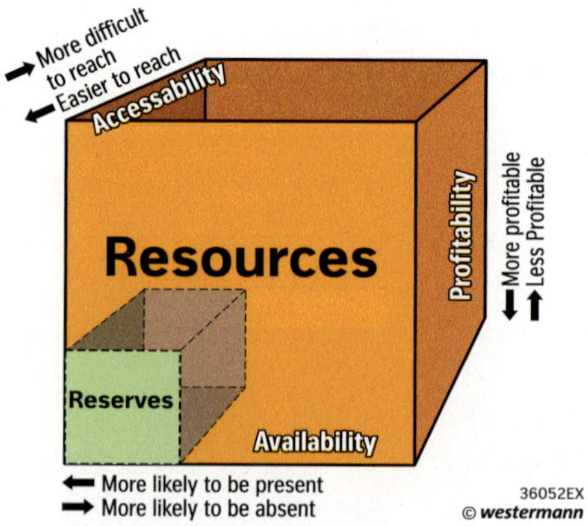

M2 Resources and reserves

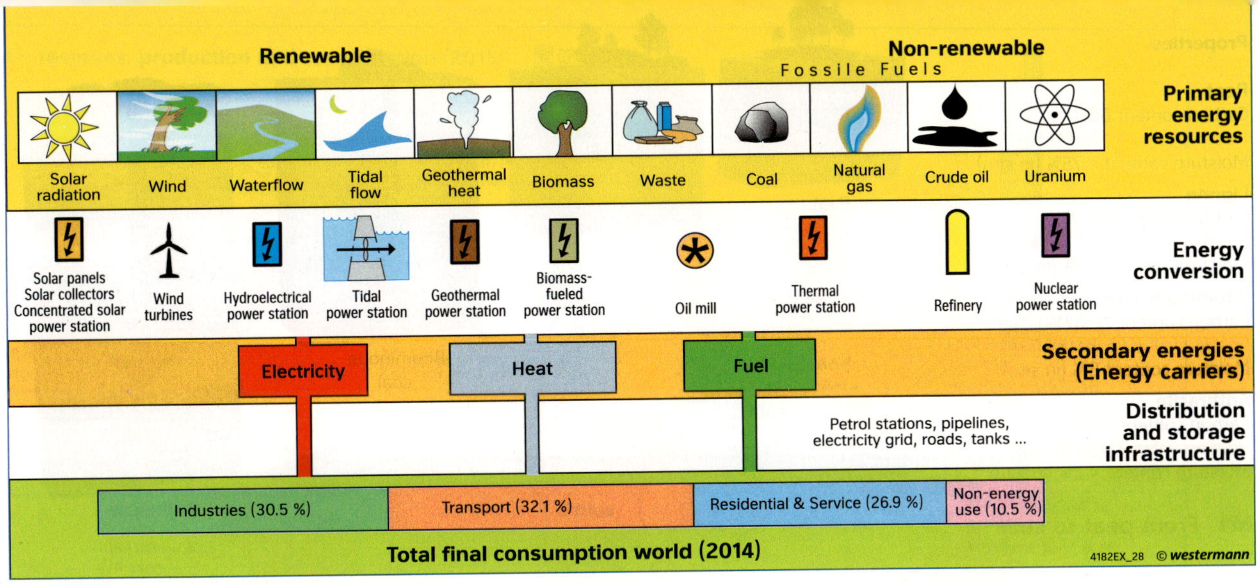

M3 Classification of energy resources

electricity. Even the energy contained in wind originates mostly from temperature differences on the globe due to solar radiation. As long as there is a sun, these resources can be used again and again. They are therefore called renewable energies. Another nearly infinite source of energy comes from the heat of the Earth's interior: geothermal energy.

A major advantage to using these renewables is that there are only few pollutants released into the environment.

Biomass is a renewable energy source not only because the energy in it comes from the sun, but also because biomass can regrow over a relatively short period of time compared with the millions of years that it takes for fossil fuels to form.

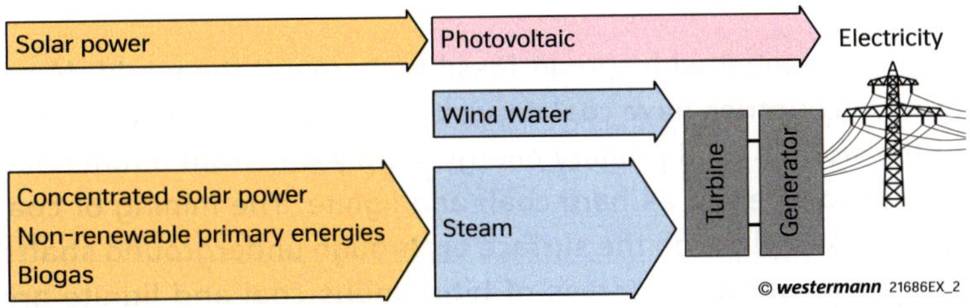

M4 Production of electricity

Primary and secondary energy resources

Primary energy is the energy within the natural resources. To use these resources, we have to convert them into energy which is fit for our use. These secondary energy resources or energy carriers can be transported to our homes and workplaces or can be used for transport.

100800-264 schueler.diercke.de

pp. 18 - 19 *Line Graphs*

TASKS

1. Summarize **M1** in four important statements.
2. Comment on the role of renewables in the world energy supply (**M1**).
3. Describe how one renewable and one non-renewable energy source makes it to the consumer (**M3**).
4. Discuss cases in which resources turn into reserves and vice versa (**M2**).

KEY TERMS

- fossil fuel
- non-renewable energy
- primary energy
- renewable energy
- reserve
- resource
- secondary energy

HELPFUL WORDS AND PHRASES

for TASK 3:

- mine
- gain (energy)
- convert
- transform
- reach
- turn into ...
- is/are distributed via ...
- is/are stored in ...
- are used for ...
- are consumed as / by ...

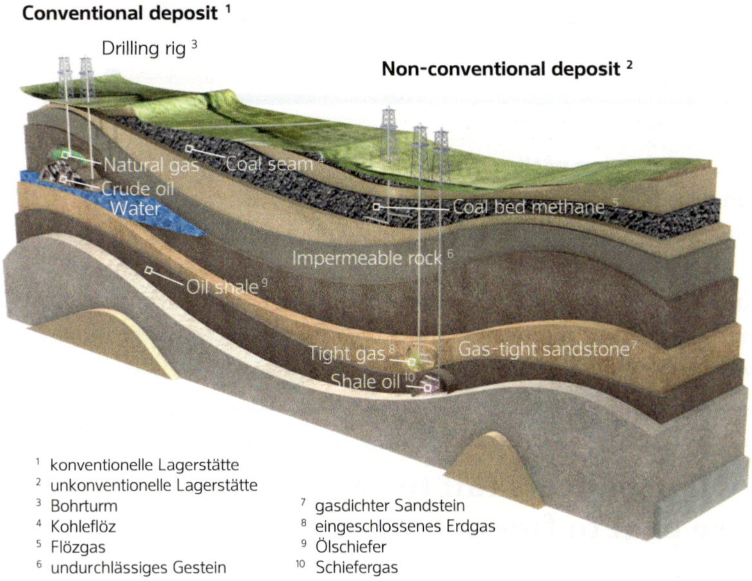

M1 Natural gas deposits

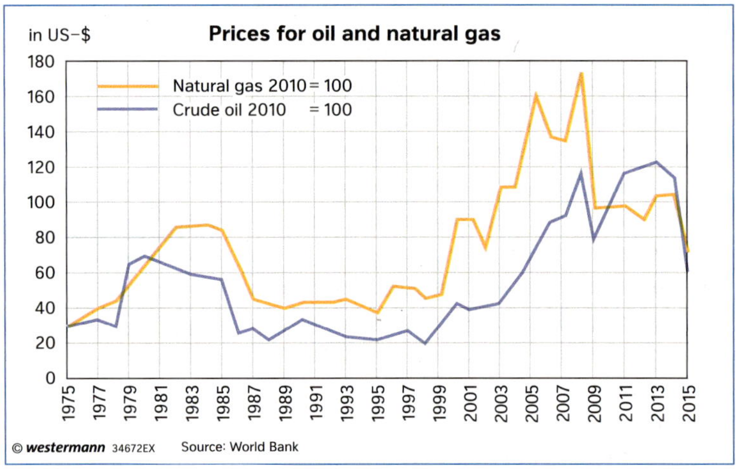

M2 World market prices for two fossil fuels

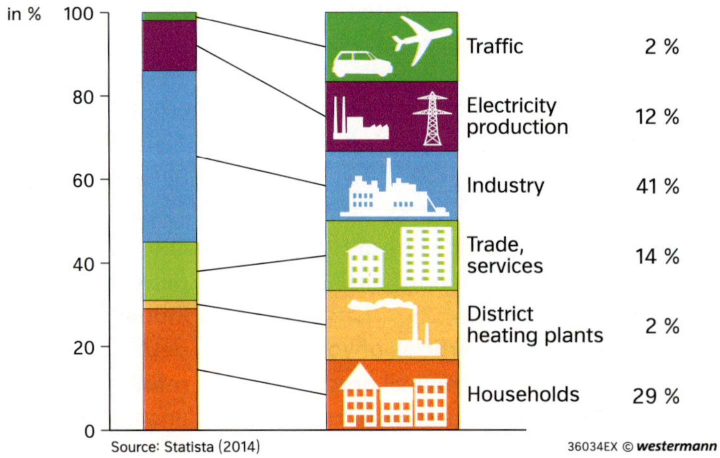

M3 Natural gas consumption in Germany

Natural Gas – A Clean Fossil Fuel?

Natural gas is a fossil fuel, mainly consisting of methane. Before the gas reaches the end consumer, it has to be processed and all impurities have to be removed. As it takes up huge volumes, it often is compressed or cooled into liquid form for storage and transport.

New techniques – increasing reserves

When oil is formed, natural gas arises as well (› pp. 60-61). Hence, gas fields and oil fields are often next to each other. Natural gas accumulates within the oil deposits and in the rocks above them.

Recently, new mining techniques have enabled access to so-called non-conventional deposits of oil and gas (**M1**). This increased the reserves and was one of the reasons that the prices for world fossil fuels dropped sharply (**M2**).

Hydraulic fracturing, or fracking, was a well-developed mining technique. When it was combined with the technique of horizontal drilling, a cost-effective way of producing natural gas (even from coal seams) became available (**M4**). Fracking immediately became the most important method of mining for natural gas.

Recently scientists found huge deposits of methane hydrate on the bottom of the oceans and in the Arctic tundra. There are, however, no recovery techniques yet to extract the methane from these deposits.

Transport

Normally natural gas is transported through pipelines. However, when it is cooled below – 161 °C, it becomes liquid, and its volume is reduced from 600 units to 1 unit. This way the liquefied natural gas (LNG) can be transported by huge tankers over long distances. Japan gets all of its natural gas through LNG tankers, whereas Germany gets all its gas through pipelines.

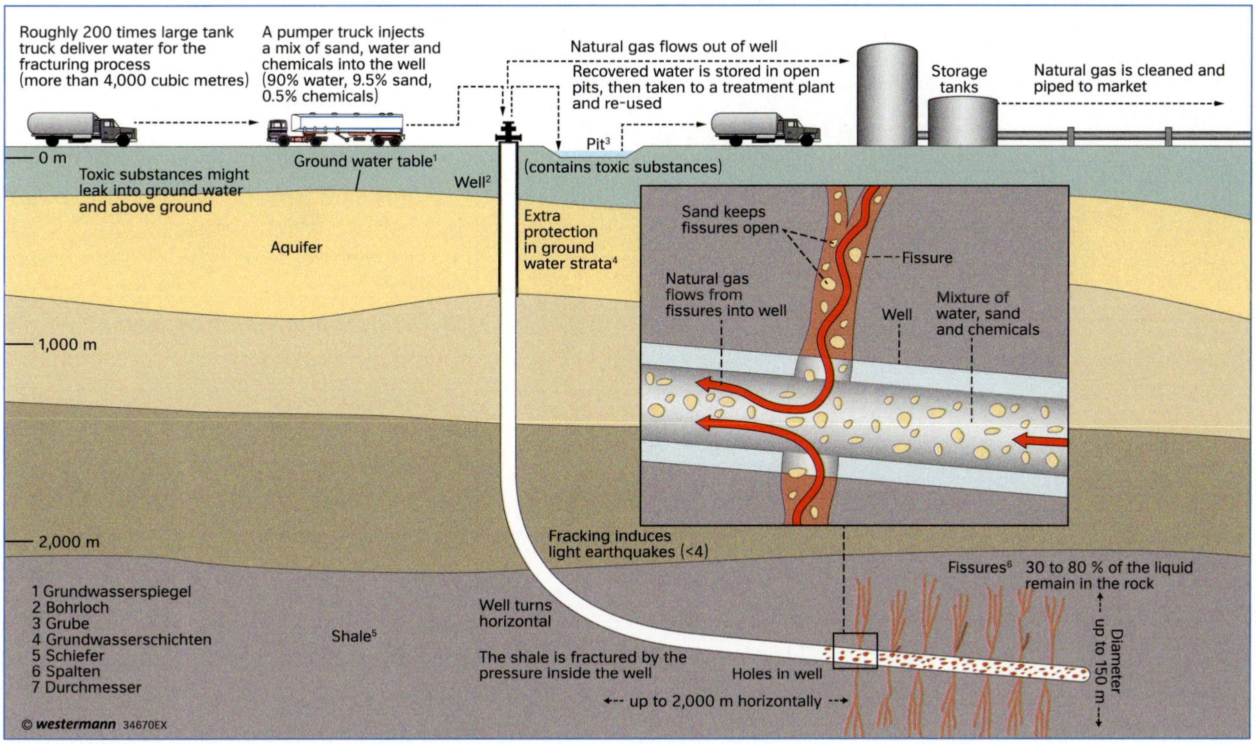

M4 Extraction of natural gas through hydraulic fracturing (fracking)

Environmental problems

Natural gas is probably the cleanest of the fossil fuels. However, mining for natural gas creates considerable amounts of drilling mud, which is poisonous and radioactive and has therefore to be disposed of carefully. In the gas processing facilities further harmful substances are removed, creating hazardous waste. The gas is flammable and leaks very easily during production, storage and transport. It is a potent greenhouse gas (GHG). Liquefied and compressed gases can expand quickly and explode. The end consumers burn the gas, creating CO_2 and water vapour which are both greenhouse gases.

Moreover, there is a danger that global warming leads to the release of trapped methane from thawing Arctic permafrost soils, contributing to the greenhouse effect.

100800-120, -220 schueler.diercke.de

DE-048, -104 www.diercke.com

TASKS

1. a) Define the unknown terms from **M1** and **M4**.
   b) Describe the mining of natural gas.
2. Comment on the similarities of oil- and gas prices (**M2**).
3. State your opinion on fracking.
4. Project in German:
   Make a poster to discuss the page title 'Natural Gas – A Clean Fossil Fuel?'

KEY TERMS

- conventional gas
- deposit
- extraction
- fracking
- liquefied natural gas (LNG)
- methane
- natural gas
- non-conventional gas
- pipeline
- tanker

HELPFUL WORDS AND PHRASES

for TASK 3:

- ... are similar/different in the following ways...
- ... are alike/differ because ...
- Likewise, ...
- In contrast, ...
- ... (do not) correspond in (size / shape / quality) ...

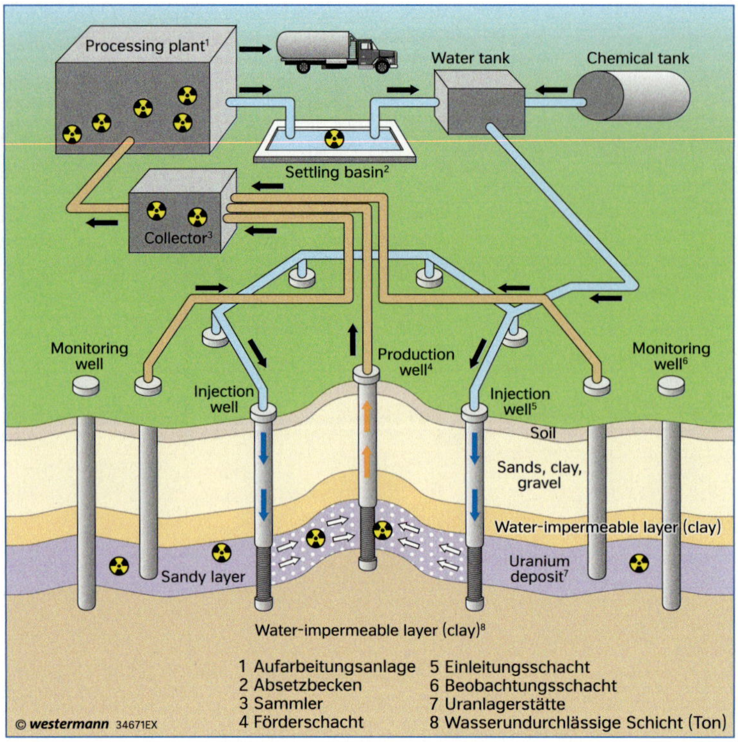

M1 In-situ leaching (ISL)

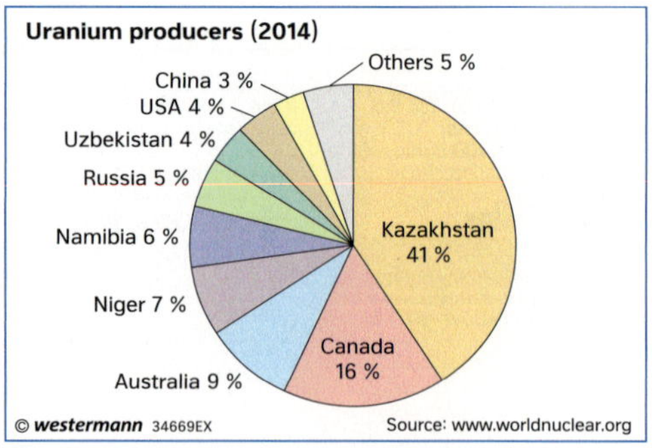

M2 World's uranium producers

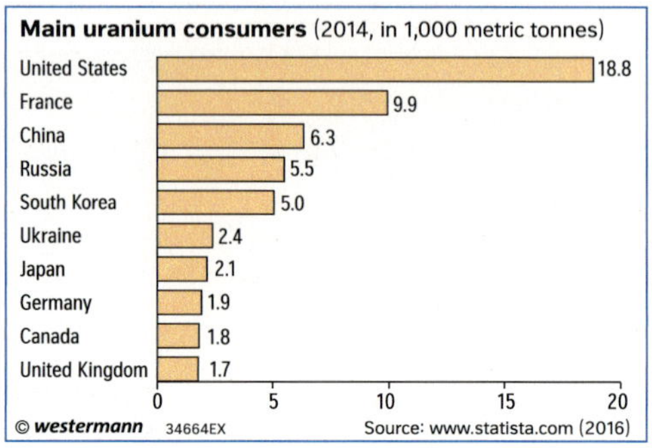

M3 World's main uranium consumers

Nuclear Energy – Friend or Foe?

100800-066, -099
schueler.diercke.de

Under certain conditions, elements can emit radioactive rays. This is always connected with the release of energy, mostly heat. When this happens, using suitable elements under controlled conditions, the heat can be used to produce electricity in so-called nuclear power stations.
Uranium is the fuel which is mostly used in these power plants. The energy in uranium is very high: 1 kg contains the same energy as 92 460 kg of crude oil.

Uranium recovery

The majority of uranium deposits is low in uranium content. Therefore, mining uranium through open pit or shaft mining creates a lot of waste materials. However, half of the world's uranium ores are recovered today through in-situ leaching (ISL, **M1**), where water and added chemicals mobilize the uranium salts in the bedrock.
Most of the world's uranium is not extracted in the countries where it is consumed.

The nuclear cycle

In a nuclear power station the fission of uranium is slowed down and thus controlled. During fission, other radioactive elements are created, most of all plutonium. In reprocessing plants the plutonium is separated from remaining uranium and both can be used for new reactor fuel (**M4**).

The nuclear debate

Nearly all the substances which are used and created in the nuclear fuel cycle are highly poisonous. Moreover, most of them are radioactive for a long time. They harm and destroy living cells. In all steps of the nuclear power cycle, it is possible that these substances could escape into the environment.
Therefore security and safety standards in nuclear facilities are extremely high. But after the nuclear disasters in Chernobyl (1986) and Fukushima (2011), concerns about using nuclear power have increased. Besides the risk of accidents, there is the

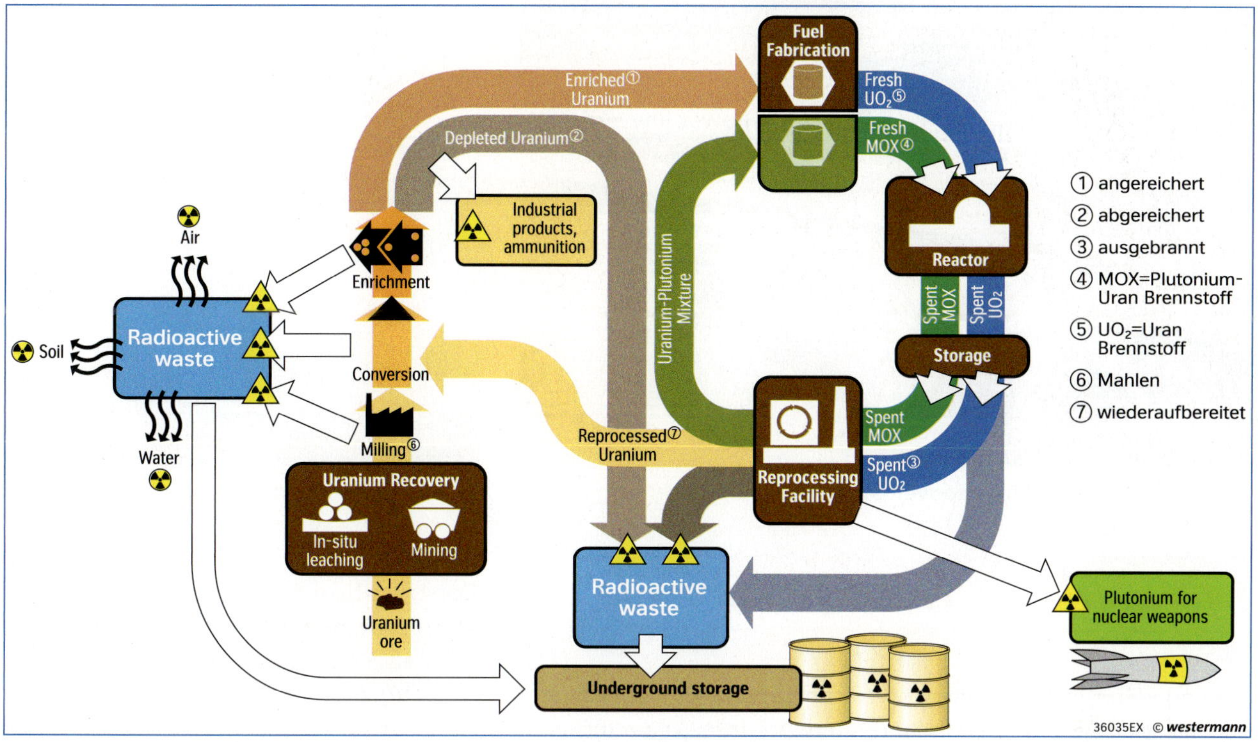

M4 The nuclear fuel cycle

unsolved problem of storage and disposal of the radioactive waste. Consequently, there are, often emotional, discussions regarding the future use of nuclear power. All of Germany's nuclear reactors will be shut down in the near future.

Other countries, however, plan to go ahead with nuclear power. They see it as a safe and inexpensive energy in which much fewer greenhouse gases are released into the atmosphere.

M5 Anti-nuclear protest (USA)

TASKS

1. a) Describe in-situ leaching (ISL, **M1**).
   b) Compare it to mining.
2. Describe the nuclear fuel cycle (**M4**).
3. Discuss the pros and cons of producing electricity by nuclear energy.
4. For experts: Prepare a speech, in German, about a nuclear accident.

KEY TERMS

- fission
- in-situ leaching
- nuclear disaster
- nuclear energy
- nuclear fuel cycle
- nuclear power station
- radioactive waste
- reprocessing facility

pp. 20-21 Bar Charts

pp. 26-27 Pie Charts

M1 Hydroelectric dam (Hoover Dam, USA)

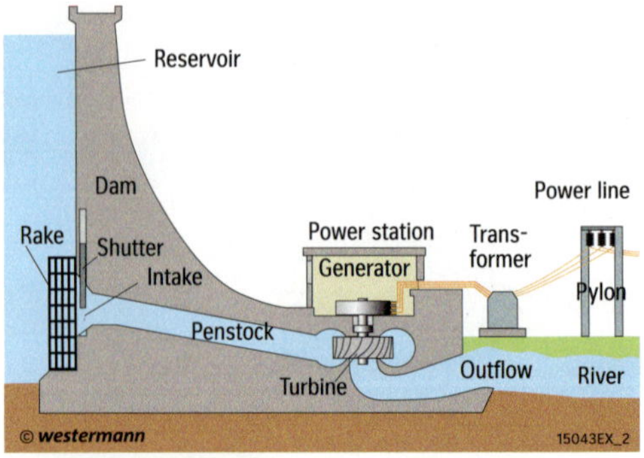

M3 Cross section of a hydroelectric dam

Hydropower – Electricity from Water

Moving water contains energy. We can capture this energy and convert it into electricity. On land, there are three main forms of doing this (**M1**, **M2**, **M5**). They all need turbines which turn generators to produce the electricity and transformers to feed the electricity into the grid.

Hydroelectricity

Mountainous regions with much rain in the catchment areas of the rivers are the preferred locations for hydroelectric dams (**M1**). Run-of-the-river power stations are often located in big rivers (**M2**). Alternatively, they can be found as small, low-tech micro-hydro power stations to produce electricity in remote regions (**M4**). In pumped-storage power stations, the water of an upper storage reservoir serves as stored energy which can be turned to electricity when needed (**M5**). They are a very important backup, especially when wind and sun do not produce sufficient electricity.

Contribution to world electricity

At present, hydropower generates approximately 17 per cent of the world's electricity (**M6**). More than three-quarters of the world's electricity from renewables comes from hydropower. Today there are four big dams in China and South America with an installed capacity of more than 10 GW each. Two more shall be completed in the near future. There are huge reserves and it is estimated that the capacity of hydropower will grow by more than 3 per cent each year in the future.

Pros and Cons

The proven and simple technology of hydropower is applicable even on a very small scale. However, a lot of capital is needed to build huge hydroelectric dams. But

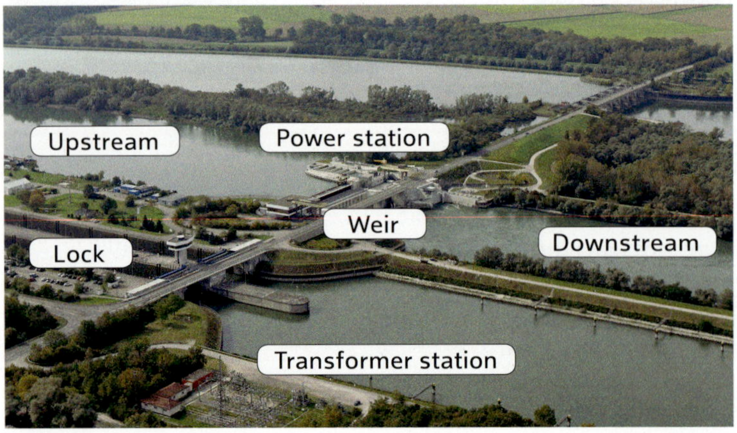

M2 Run-of-the-river power station (Rhine, Germany/France)

M4 Micro-hydro installation

Energy

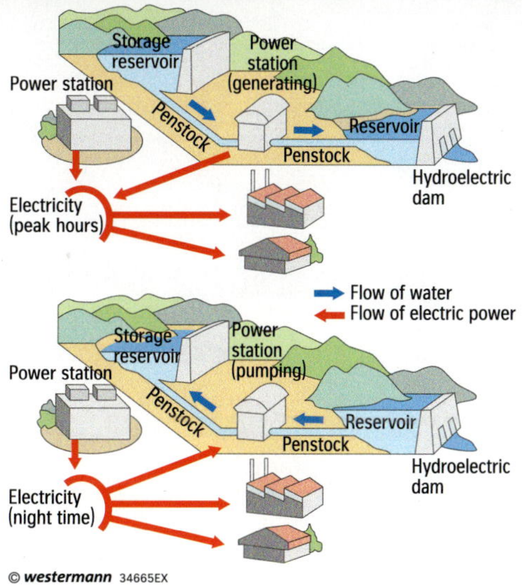

M5 Model of a pumped-storage power station

Producers	Output (TWh)	Amount (%) of world total	Amount (%) of country's electricity
China	1,064	26.7	18.8
Canada	383	9.6	58.4
Brazil	373	9.4	63.1
United States	282	7.1	6.6
Russia	177	4.4	16.7
Rest of the World	1,427	42.8	14.2
World Total	3,874	100.0	16.7

Includes electricity production from pumped storage; Source: OECD/IDEA (2016)

M6 Main producers of hydroelectricity (2014)

once the project stands, the operating costs are low, making the production costs per kWh the lowest of all energies.
The big dams and their lakes use up a lot of land, so that resettlement become necessary. Research has shown that hydroelectric dams change the microclimate and much water evaporates. On the other hand, no greenhouse gases (GHGs) and no waste are created. Also, after the energy has been yielded, the water is still there for other purposes in agriculture, industry, households, and thermal power stations.

El Niño

El Niño is a climatic event which occurs irregularly every two to seven years. Pacific surface waters warm up, and the upwelling of cold, nutrient-rich water is reduced off the western coast of South America. This influences weather patterns, ocean conditions, and marine fisheries and affects the climate across large parts of the world.

The effect of drought on Brazil's hydropower

Brazil, the world's second largest producer of hydroelectricity, is currently facing its worst drought in 40 years. The lack of rainfall caused hydropower output to fall by 7% in 2013.
2014 was another dry year for the country, with its largest reservoirs only at 16.1% of capacity on average at the year's end. Brazil's government has been forced to turn to more costly thermoelectric plants in order to avoid blackouts. Industrial customers [...] now are facing prices that are more than double the amount customers paid as of January 2014.
Though El Niño was expected to increase the amount of rainfall over the winter, changes in weather patterns in Brazil resulted in El Niño not travelling far enough to fill key reservoirs in the north of the country. [...]

Source: www.worldenergy.org (World Energy Resources: Charting the Upsurge in Hydropower Development, 2015)

M7 ◁ Hydropower and climate change

100800-066, -189 schueler.diercke.de

DE-110 www.diercke.com

pp. 12 - 13 *Photographs*

pp. 16 - 17 *Tables*

KEY TERMS

- hydropower
- hydroelectric dam
- micro-hydropower station
- pumped-storage power station
- run-of-the-river power station
- turbine

TASKS

1. Compare hydroelectric dams with run-of-the-river power stations (**M1**, **M2**).
2. Explain the workings and the importance of pumped-storage power stations (**M5**).
3. Comment on the effect of drought on hydropower (**M7**).

M1 An open fire for cooking and heating

M3 Maize silage harvest for a biogas plant

Bioenergy – Not Just Firewood

100800-068, -069
schueler.diercke.de

Renewable bioenergy

Bioenergy is the largest renewable energy source today. It is a renewable energy source not only because the energy in it comes from the sun, but also because the energy-rich biomass can regrow over a relatively short period of time compared with the hundreds of millions of years that it took for fossil fuels to form.
Through the process of photosynthesis, plants capture the sun's energy by converting carbon dioxide from the air and water and minerals from the ground into energy-rich plant matter and oxygen. When the biomass is burned, carbon dioxide, water, and the energy are released. It is a cycle, driven by the sun.

Traditional use

In India alone, 160 million households use wood, charcoal, agricultural residues and animal dung for cooking and heating. This normally has a very low conversion efficiency: only 10 to 20 per cent of the available energy is actually used. The supply is often not sustainable as the resources of firewood are depleted. Moreover, severe health and safety problems arise from the smoke and the open fires (**M1**).

The role of bioenergy today

Over the last decade, bioenergy has become increasingly important as a substitute for fossil fuels. Today, more than 10 per cent of the world's primary energy supply comes from the conversion of biomass, including waste. This energy is mainly used for cooking and heating (in EU 2013: 75 per cent), but also to make electricity (in EU 2013: 13 per cent) and to produce fuel for transport (in EU 2013: 12 per cent, **M2**).

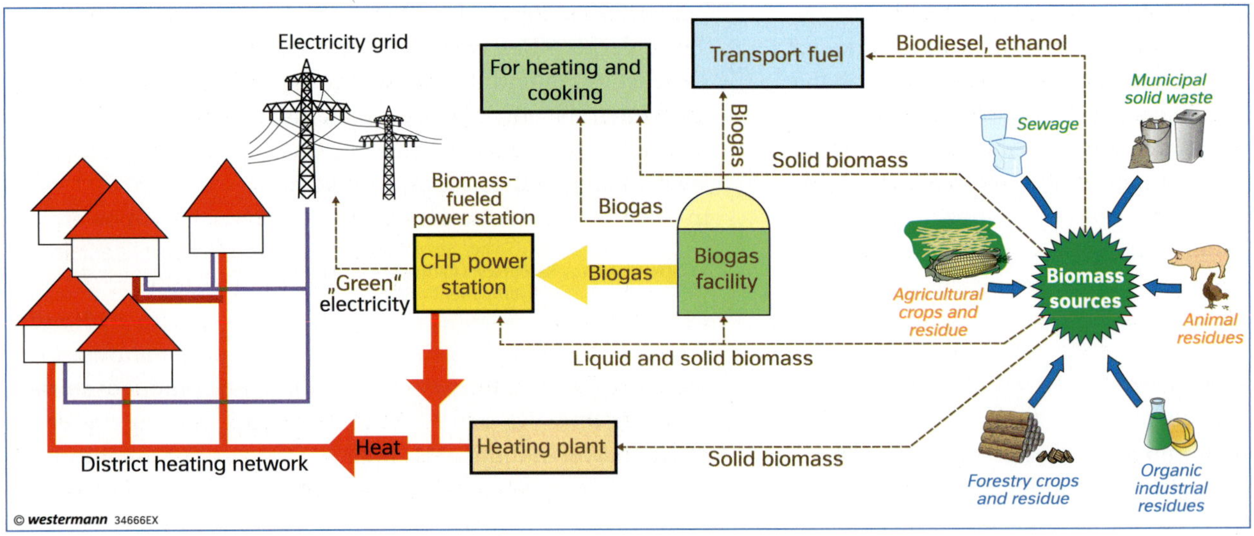

M2 Biomass – a renewable energy source

The prospects of bioenergy

Biomass can be grown almost everywhere, so bioenergy can be produced where and when it is needed. When it is used sustainably it can be regrown every year (› p. 94). Biomass can be turned into whichever form of secondary energy is in demand. To do this effectively however, technical and chemical know-how is essential. Energy assessments are necessary to find the right energy conversion, as the energy yield not only depends on the crop, but also on the technology involved (**M4**).

Especially in less economically developed countries (LEDCs) where the agricultural land and water are scarce, these resources are often used to grow energy crops rather than food crops, even when the rural population is starving.

Moreover, the growth of energy crops is often extended into areas which are not suitable for farming. This can lead to soil degradation, erosion, and the destruction of pristine landscapes (**M5**).

However, new technologies and different forms of biomass production will increase the effective and sustainable use of clean bioenergy in future.

TASKS

1. Describe biomass as a renewable energy source (**M2**, text).
2. a) Prepare five questions about the current and future role of biomass as an energy source.
   b) Discuss the questions in class.
3. Compare the advantages and disadvantages of ethanol as a biofuel (**M5**, text).

KEY TERMS

- bioenergy
- biofuel
- biogas
- biomass
- energy conversion
- energy crop

Energy source (residues, energy crops)	Conversion process	Yield*
Forest waste wood	Combustion	434
Straw	Combustion	2,390
Maize silage	Biogas	5,280
Rapeseed oil	Combustion, biodiesel	1,528
Short-rotation coppice trees	Combustion	5,120
Whole-plant grain silage	Biogas	4,013
Grain kernels	Combustion, biogas, ethanol	2,232
Grass silage	Biogas	3,016
Miscanthus grass, dried	Combustion	6,081

*Heating oil equivalent in l/(ha x a) [a tank lorry can carry 22,000 litres of heating oil]; Source: www.bioenergie.fnr.de (2014)

M4 Energy content of biomass sources

Ethanol from corn and sugarcane

Corn requires large doses of herbicide and nitrogen fertilizer and can cause more soil erosion than any other crop. And producing corn ethanol consumes just about as much fossil fuel as the ethanol itself replaces. [...] Environmentalists fear that rising prices (for corn) will push farmers to plow up [...] marginal farmland now set aside for soil and wildlife conservation, potentially releasing even more carbon bound in the fallow fields. The boom has already pushed corn prices to heights not seen in years. [...] Around a fifth of the harvest will be brewed into ethanol. [...]

Thirty years after launching a crash program to replace gasoline with ethanol from sugarcane, Brazil announced last year that thanks to ethanol [...] it had weaned itself off imported oil. [...] 'We get eight units of ethanol for every one unit of fossil fuel', says [...] one of Brazil's leading sugarcane researchers. Experts estimate that producing and burning cane ethanol generates anywhere from 55 to 90 percent less carbon dioxide than gasoline. [...] The expansion of Brazil's cane acreage ... may also be contributing to deforestation. By displacing ranching in existing agricultural areas, sugar may be adding to the pressures that send cattlemen deeper into frontier territory like the Amazon and the [...] savannas. A recent UN report concludes that [...] the biofuels boom could reduce food security and drive up food prices in a world where 25,000 people die of hunger every day. [...] Demand for both fuel and food is expected to more than double by mid-century, and many scientists fear that in coming decades, climate change will undermine agricultural productivity.

Source: www.nationalgeographic.com (Joel K. Bourne, 'Green Dreams', 18 May 2016)

M5 Excerpts from a magazine article

Solar Energy – Let the Sunshine In

Latitude, season, time of day, cloud cover, pollutants in the atmosphere, height above sea level: all of these factors affect how much sunlight reaches the earth's surface (**M3**).
Light can be used to generate electricity or can serve as a heat source. The appliances (**M2**) are plentiful, as the source sunlight is ubiquitous, free, renewable, and does not produce waste.
Costs for the appliances have decreased over the years so that more and more of them are installed worldwide. Small applications can be particularly useful in regions where people have no access to the electric grid.

M1 Solar farm in Germany

Earning money with solar panels
Solar cells can turn light to electricity through photovoltaic (PV). These cells can be connected to make larger solar panels which can be arranged in solar farms. More frequently, solar panels are used on buildings to supply them with electricity. When there is more power produced than needed, it can either be stored in batteries or fed into the grid for money.

Hot water for free
We need a considerable amount of hot water each year, be it in bath and kitchen or for heating.
Therefore, many households use rooftop solar collectors to make hot water and in this way reduce their heating costs.
Additionally, this reduces the need to burn fossil fuels which are normally used for heating.

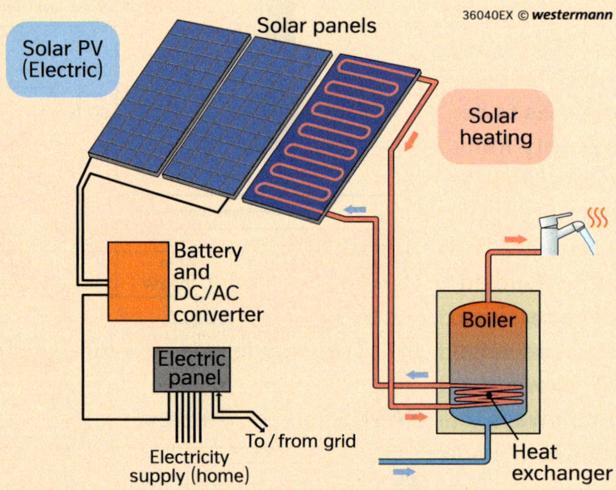

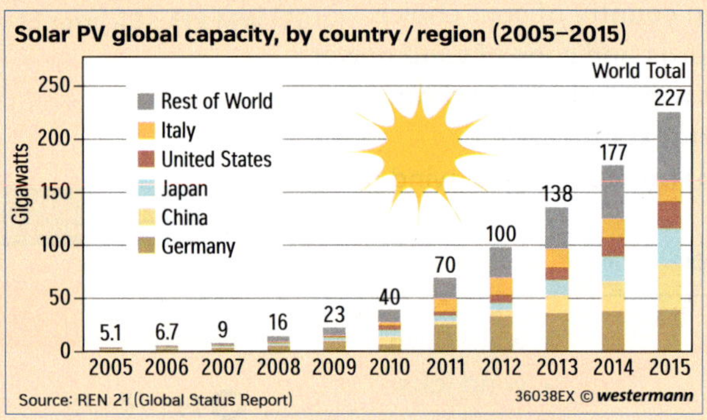

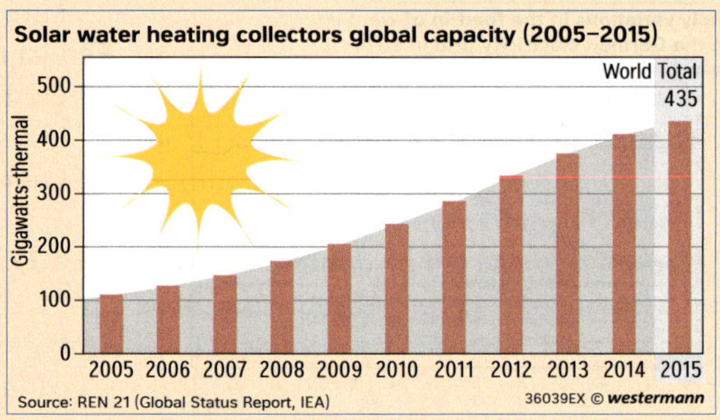

M2 The most important applications to use sunlight as an energy source

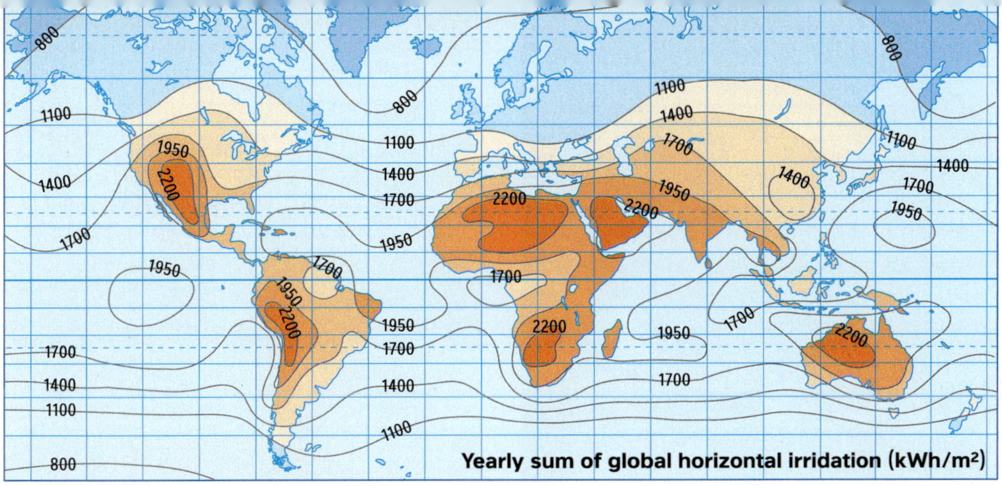

M3 Global sunshine power

Concentrated solar power

Concentrating sunlight can create high temperatures. For areas with little biomass fuel available, solar cookers have been developed. Food can be prepared and water can be distilled with no energy costs.

In many areas of the world different concentrated solar power (CSP) technologies have been developed to drive thermal power plants with this heat. Installation costs are still very high. It is thought that they are especially well suited for deserts and semideserts. Much energy is needed there to cool homes and public spaces.

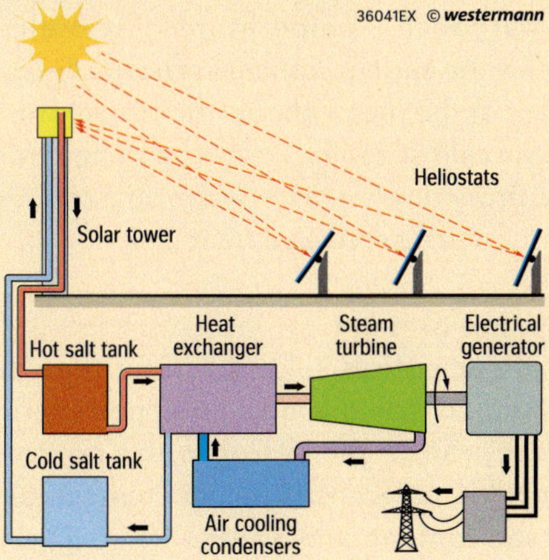

Supply and demand

The time for electricity demand not always correlates with the time of supply of sun and wind energy. Not to lose too much energy, storage facilities turn the electricity into other energies like gas for further use (› p. 135: **M5**).

Energy conversions and storage always lead to energy losses. To minimize these losses is an ongoing task.

100800-068, -069
schueler.diercke.de

pp. 18 - 19
Line Graphs

pp. 20 - 21
Bar Charts

TASKS

1. Compare in a table the uses of solar energy shown on these pages.
2. Create a 45-second TV commercial / an advertising poster for solar panels.
3. For experts: Discuss the use of solar power in LEDCs and MEDCs.

HELPFUL WORDS AND PHRASES

for TASK 1:

- ... in the same way.
- ... in relation to ...
- ... look (un)like ...
- In contrast, ...
- On the one hand, ... on the other hand, ...

KEY TERMS

- concentrated solar power (CSP)
- photovoltaic (PV)
- solar cell
- solar collector
- solar energy
- solar panel
- solar tower

M1 Geothermal power station in Puhagan (Philippines)

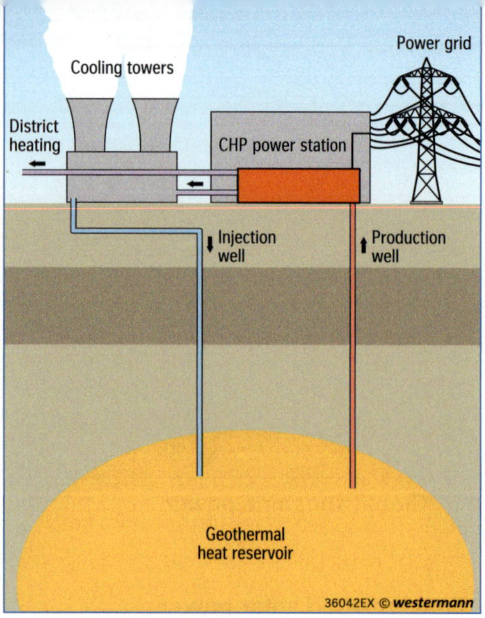

M2 Model of a geothermal power station

Other Renewable Energies

Geothermal energy

The heat of the earth's interior is a huge energy reservoir. Away from plate boundaries, the temperature of the lithosphere increases by about 25 °C per km. Under certain geological conditions, heat reservoirs may be much nearer to the surface. They can be exploited for district heating or electricity production (**M1**, **M2**). Although initial investments are often high, running costs and pollution are quite low. Research suggests that the overall costs per kWh can be even lower than the cost for wind energy.

Heat pumps

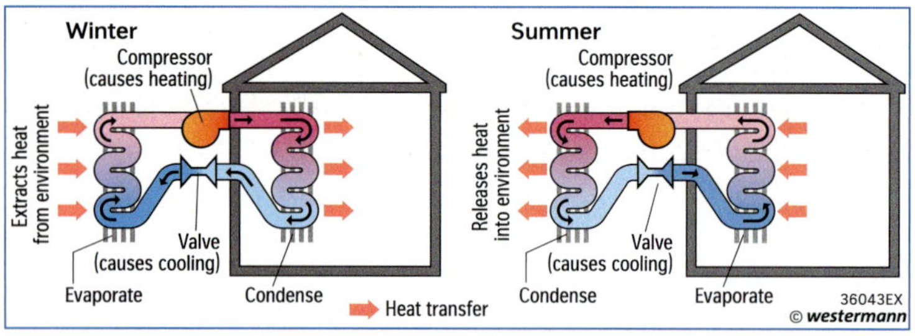

Heat pumps work like refrigerators. They can cool, for example, warehouses, lorries, homes, and offices by pumping heat energy to the outside. Equally important is, however, that the pumps can work in the opposite direction, too. They then heat the buildings by extracting thermal energy from outside. Heat pumps run on electricity but they need much less energy than they gain. Between 2009 and 2015, in Germany alone, around 400,000 heat pumps had been installed.

Ambient heat

Even relatively small temperature differences between the surface and the environment are sufficient to heat or even cool homes, offices and workshops with the help of heat pumps.

Power companies have begun to use this so-called ambient heat (heat from the environment) to provide district heating and cooling. Rivers, lakes, oceans, the ground, and air – in fact everything emitting heat – can be tapped for energy (**M3**).

Temperature changes of the environment are kept to a minimum to avoid ecological disturbance. Temperatures six metres below the earth's surface are normally constant at the mean annual air temperature. In the cold seasons, heat is extracted from the ground, whereas in warm seasons, the ground can serve as a heat sink.

Marine energy

Many scientists take the view that the oceans contain the most untapped energy. Energy can be gained from ocean currents, the tides, differences in water temperature or salt content, and waves.

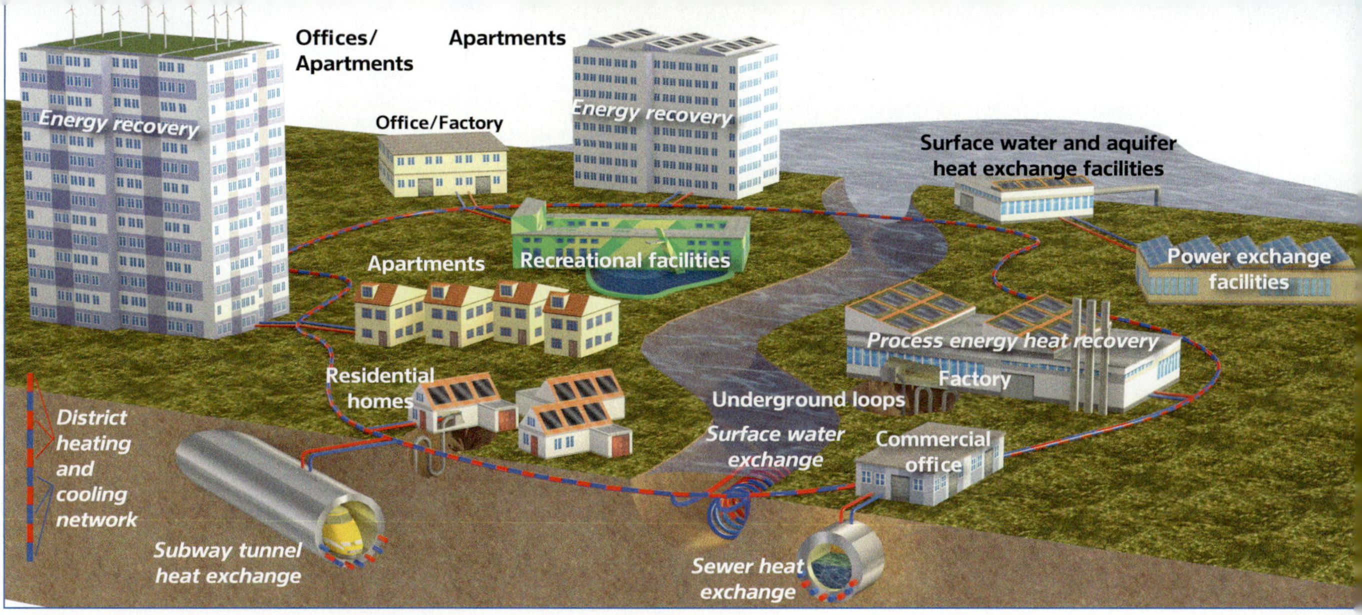

M3 Sources and uses of ambient heat

Tidal energy

You can use the rise and fall of the tides to generate electricity, especially where the tidal range is high. Unfortunately, there are not many suitable locations on Earth. The building costs are normally quite high, but once the tidal power station is in operation, it provides free energy with no pollution and low maintenance. One important aspect is that tidal energy is a rather predictable energy supply.

On the other hand, there are disturbances to coastal shipping, since the dams serve as bridges. The environmental impact on the tidal flats can be great although modern tidal barrages pose no problems for fish and other creatures to cross. The areas behind the dams often become recreational lakes and attract tourists.

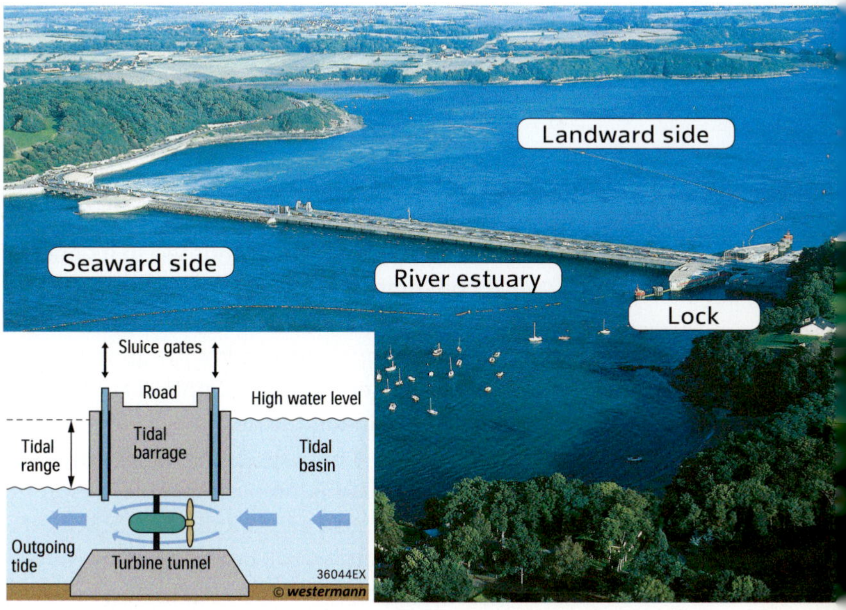

M4 Tidal power station at Rance (France) and model of a tidal power station

TASKS

1. Describe the working of the two power stations shown with the help of their models.
2. Write a text about the many sources and uses of ambient heat (text, **M3**).
3. For experts: Research for a presentation in German about one of the marine energies.

KEY TERMS

- ambient heat
- geothermal energy
- heat pump
- marine energy
- tidal barrage
- tidal energy

100800-068, -108
schueler.diercke.de

DE-049
www.diercke.com

pp. 12 - 13 Photographs

pp. 32 - 33 Texts

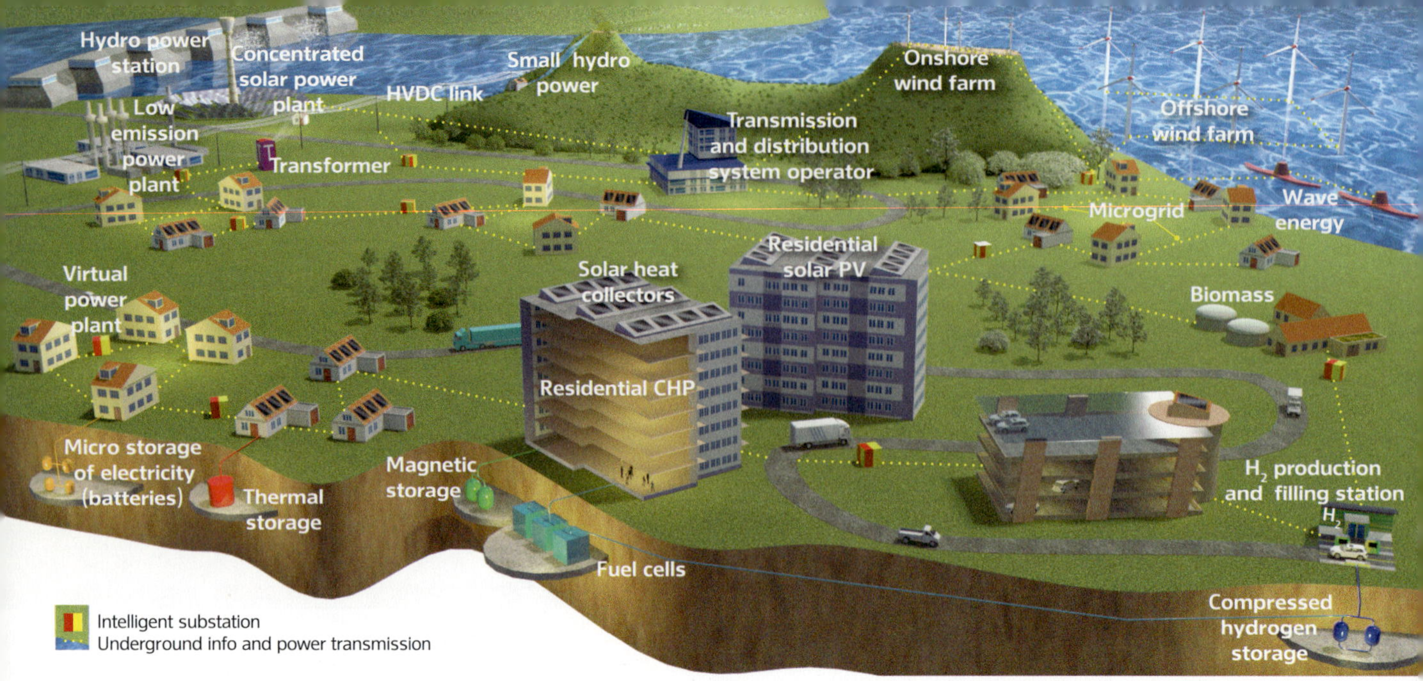

M1 Smart grid of the future

World Energy – Challenges and Perspectives

Energy

The development of a country's economy is always very closely related to its energy consumption per capita. As the world's population continues to grow and the level of development worldwide is meant to increase, energy consumption is very likely to rise in the future.
However, a big portion of the energy produced today is non-renewable, resulting in greenhouse gases and other pollutants. These energies will be replaced by renewable energies amid growing energy demands.

MEPS

Minimum energy performance standards (MEPS) are often part of governmental policies. The European Union's 'Ecodesign Directive' regulates the use of lightbulbs, boilers, TVs, and fridges for example to reduce energy consumption. Thus, the EU claims, the annual CO_2 emissions will be reduced by 16 million tons by 2020.

Main developments

Electricity will become the major source of energy. This will require improvements in battery storage and smart grid networks (**M1**).
Decarbonisation is the reduction of the use of fossil fuels. This reduction will minimise the release of greenhouse gases (GHGs) and other pollutants. Non-organic sources, however, produce electricity and heat only. There has to be a significant increase in the use of renewables. Installing renewable energy farms on a large scale is very costly and therefore not suitable for poor countries.
Grid extensions will be necessary to transport the electricity to the consumers. However, there are many small-scale appliances available which are inexpensive and well suited for use in less economically developed countries (LEDCs).
Decentralisation and virtual power plants (**M1**) will prevent heat and electricity loss by transport. Having many small electricity producers rather than few big ones will reduce the risk of disturbances in the energy supply.

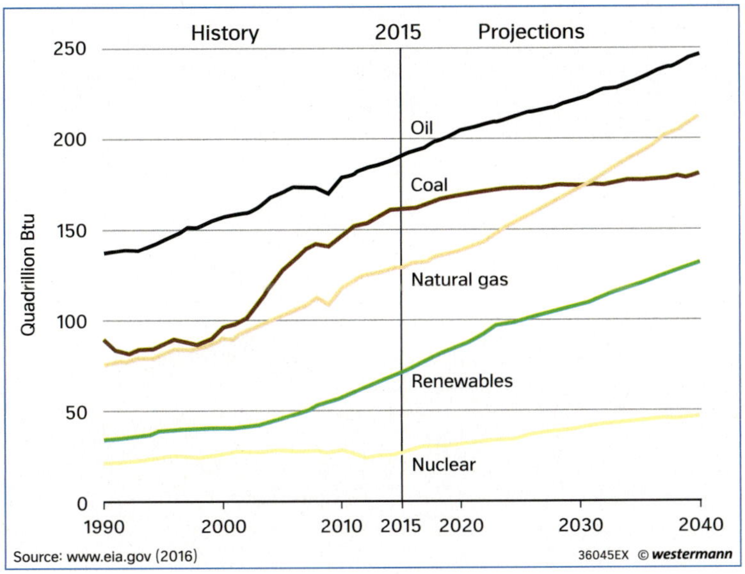

M2 World energy consumption by fuel

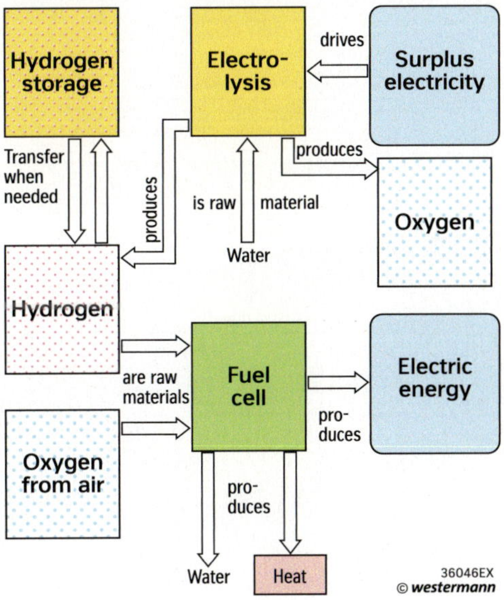

M3 Electricity, hydrogen and fuel cells

Increasing energy efficiency means that better use is made of the energy produced. This reduces the emission of greenhouse gases and saves money in energy costs for the consumers.

Governments play a significant role when it comes to the use of energy, for example through regulation, taxes and subsidies. Moreover, they can sign international treaties to agree on certain environmental goals, like the Paris Agreement of 2015 to reduce greenhouse gas emissions. Parliaments can pass laws which forbid or encourage the use of specific energy-consuming devices (MEPS).

Fuel cells

Energy for a safe electricity supply has always to be produced in excess. One way to store the surplus electricity can be to produce energy-rich gases like hydrogen (Power-to-gas, › p. 135: **M5**).

Hydrogen can be stored and transported. It can be used in fuel cells to produce electricity when needed (**M3**). The end product, water, is not harmful to the environment.

M4 Fuel cell bus in London

pp. 18 - 19 Line Graphs

HELPFUL WORDS AND PHRASES

for TASK 1:

- How does ... work?
- In what way are ... and ... connected.
- Can you explain ...
- Can you tell me more about ...

TASKS

1. a) Prepare five questions about smart grids (**M1**).
   b) Discuss them in class.
2. Explain the challenges of future energy use (text, **M2**, **M1**).
3. Make a list of things you could do to use energy more efficiently.

KEY TERMS

- decarbonisation
- decentralisation
- energy efficiency
- fuel cell
- smart grid
- virtual power plant

Ice bergs
Potato harvest
Haulm

Environment – Global Change

80 Changing Climate
82 People Leave Their Carbon Footprint
84 Soil Degradation – Losing Arable Land
86 Fresh Water – A Vital Resource
88 Threatened Oceans
90 The Ecological Footprint

M1 Due to climate change, it is now possible to grow potatoes on the southern coast of Greenland

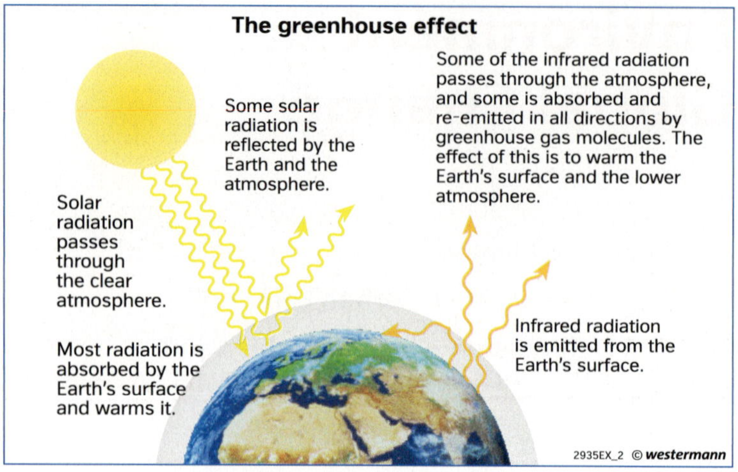

M1 The greenhouse effect

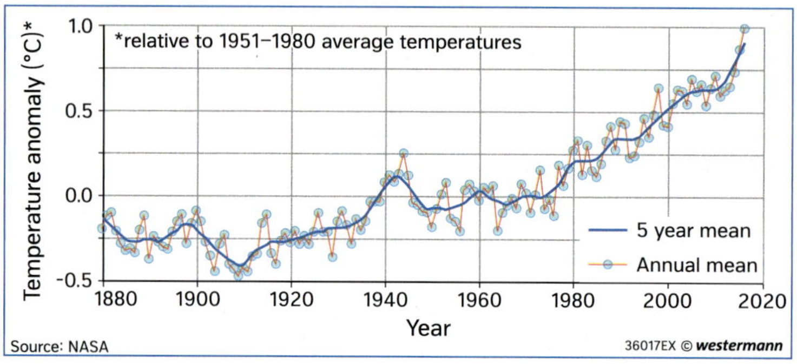

M2 Global average temperatures

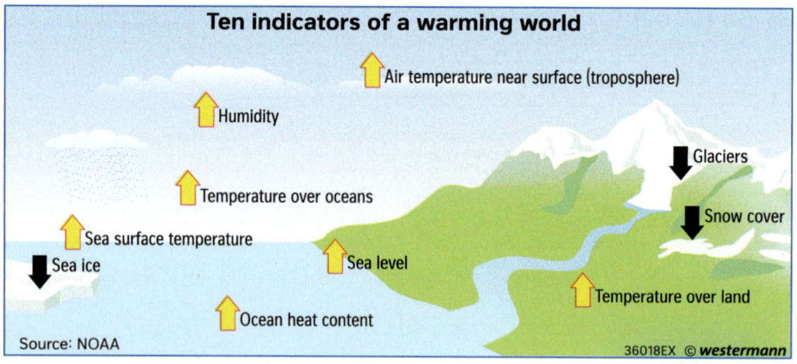

M3 Impacts of global warming

GHG	Chemical formula	GWP_{100}
Carbon dioxide	CO_2	1*
Methane	CH_4	28–36
Nitrous oxide	N_2O	265–298
Flourinated gases	(several)	124 – 22 800

*used as the reference for a 100-year timescale; Source: IPCC (2007), EPA (2016)

M4 Global warming potentials (GWP) of some greenhouse gases (GHGs)

Changing Climate

From analysing glaciers, ice cores, tree rings and ocean sediments, scientists know that climate has changed throughout Earth's history. There were periods of warming and periods of cooling.
Recent climate history is marked by an average increase in temperature of the earth's atmosphere of about 0.8 °C within the last 100 years (**M2**). This process is often called global warming.

The greenhouse effect

The amount of solar energy which reaches the Earth is, theoretically, sufficient to produce an average surface temperature of -18 °C. However, measurements have shown that the actual average temperature of our planet is about 15 °C. The 33 °C difference is caused by a natural process called the greenhouse effect (**M1**).

Global warming

Within the last several decades, scientists have been involved in a very controversial discussion on the causes of global warming. The overwhelming majority of climate scientists hold the view that a man-made increase in greenhouse gases might heat up the atmosphere by about 5 °C by the year 2100.
Since the beginning of industrialisation, the natural balance of greenhouse gases (GHGs) in the atmosphere has undergone a change caused by various human activities. Burning fossil fuels releases CO_2, CH_4 and water vapour. The decomposition of vegetation (in the absence of oxygen), livestock farming, as well as the cultivation of rice contribute to an increase in the amount of greenhouse gases. Winds spread the GHGs throughout the atmosphere leading to a globalisation of this increased greenhouse effect.
Scientists fear, that global warming will have complex effects on all living creatures (**M3**).

M5 Pasterze glacier (Austria) in 1938

M6 Pasterze glacier (Austria) in 2013

Climate change in the Alps

People do not recognize changes in global mean temperatures. However, they can observe the local impact these have. In mountain regions, like the Alps for example, the effects are striking. Since 1901, the annual mean temperature in Germany has increased by 0.9 °C, whereas the increase in the Alps is almost 1.5 °C.

Based on computer-simulated climate models, scientists have predicted the following changes for the Alps region:
- a decrease in the number of days with frost,
- more rainfall than snow,
- and an increasing risk of extreme weather conditions.

At the end of the 20th century, the total area and the mass of glaciers was only half of that in 1850. Since a large number of glaciers in the Alps cover an area of less than 1 km^2, probably most glaciers will disappear within the next couple of decades (**M5**, **M6**). Also, there might be an increase of risks like avalanches and landslides. Furthermore, rising temperatures will have an effect on some species of cold-adapted alpine flora.
As a result of warmer climate, the number of snow-reliable winter sport areas will be reduced, which will have economic consequences.

Conventions on climate change

Climate change is a global challenge requiring the implementation of global climate protection measures. The UN has organised international meetings in order to develop them. The latest took place in December 2015 in Paris, where a climate agreement was signed by 175 nations.
In the final document the members agreed to reduce their carbon output to keep global warming below 2 °C. This agreement will not become binding on its member states until 55 parties who produce over 55 per cent of the world's greenhouse gases have ratified it.

100800-250 schueler.diercke.de

DE-198 www.diercke.com

pp. 18-19 *Line Graphs*

HELPFUL WORDS AND PHRASES
for TASK 2:
- The global temperatures ...
  ... change constantly
  ... declined
  ... increased
  ... reached a peak / maximum
  ... reached a low / minimum

TASKS

1. Describe the greenhouse effect (**M1**). Answer in English or in German.
2. Explain the changing global temperatures during the past decades.
3. Prepare a report about possible consequences of climate change in the Alps.

KEY TERMS

- climate agreement
- climate change
- climate protection
- global warming
- greenhouse effect
- greenhouse gas

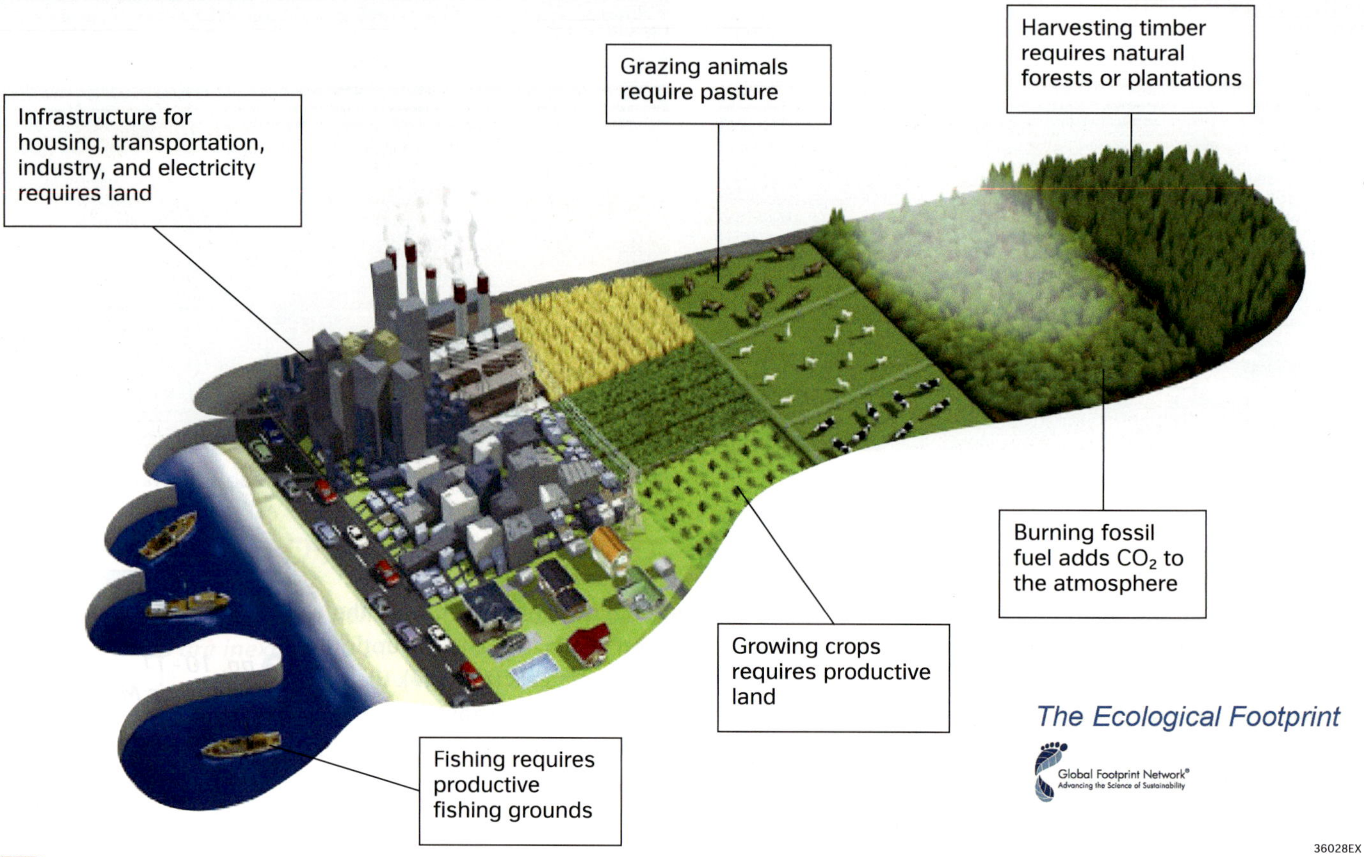

M1 People leave their marks

The Ecological Footprint

Biocapacity

In order to ensure its existence, human society has to use different natural resources. All these activities have an impact on various ecosystems. The capacity of an ecosystem to produce all the goods used by humans and to absorb waste materials is named biological capacity or, in short, biocapacity. It is calculated for all biologically productive land and sea areas of the world. Non-productive open oceans are excluded.

The ecological footprint

It is possible to calculate the area of biologically productive land and water that a population uses to generate the resources it consumes and absorb its waste under prevailing technology. The result is called the ecological footprint (EF) (**M1**).

For individuals this measure answers the question, how much area is necessary to support the lifestyle of a person. In the same manner it is possible to calculate such a footprint for towns, countries, or for humanity as a whole.

When the EF of a city or a country exceeds the biocapacity, the consequence is an ecological deficit (**M2**). An ecological reserve exists when the biocapacity exceeds the footprint.

The ecological overshoot

As long as the use of natural resources does not exceed what the Earth can supply or regenerate, there is no need for concern. However, when the demand surpasses the Earth's capacity, we move into what is named the ecological overshoot. Examples are the growing demand and depletion of non-renewable resources such as crude oil, natural gas, and minerals and the pollution of renewable ones.

M2 Ecological footprint of Berlin

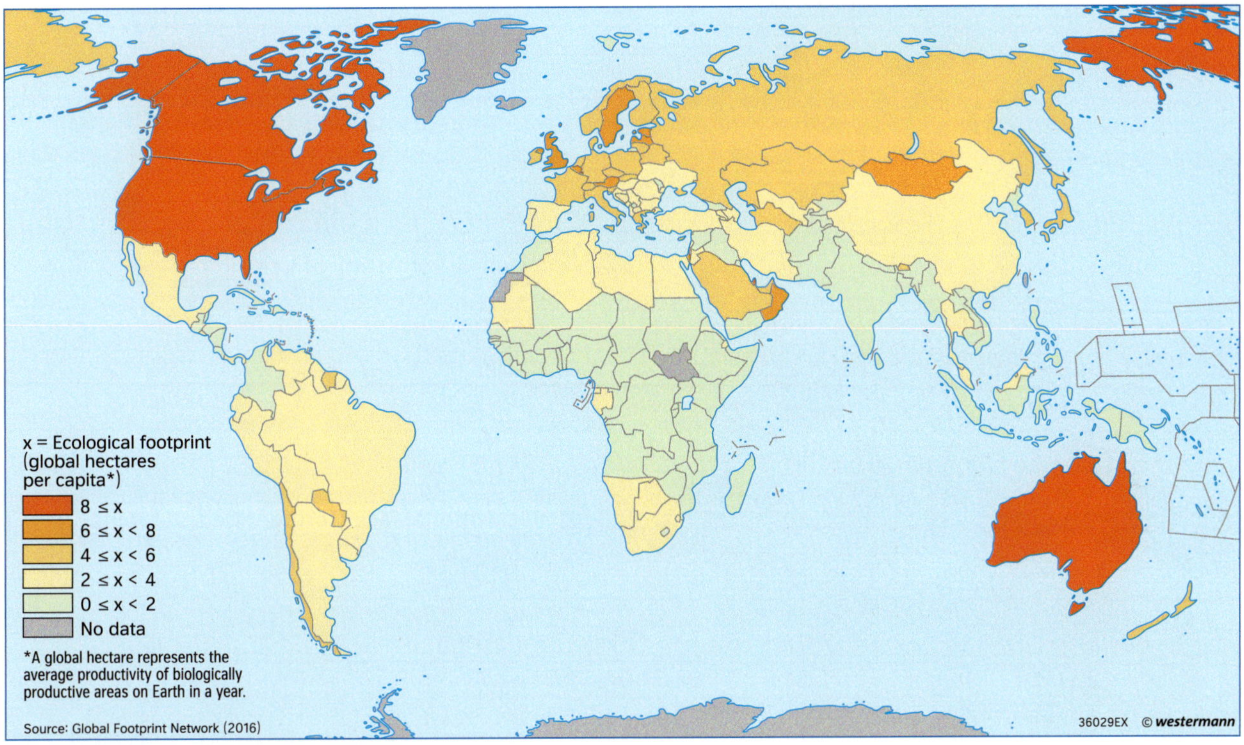

M3 Ecological Footprint of countries (2012)

At present, it is estimated that the global Ecological Footprint exceeds the earth's biocapacity by 25 per cent. This ecological overshoot affects more economically developed countries (MEDCs) and less economically developed countries (LEDCs) alike.
The so-called Earth Overshoot Day (EOD) is the date on which humanity's consumption of different resources for the year exceeds the earth's capacity to regenerate those resources in that year.

Shrink and Share

Shrink means reducing the ecological footprint to the point where the consumption of resources does not exceed the Earth's ability to regenerate. Share refers to the way the world's biologically productive capacity is divided among individuals, nations and regions.
A concerted effort by all countries and people is necessary to reverse growth in the ecological footprint and to prevent the global ecosystem's eventual collapse. Some countries will need to reduce their consumption of natural resources to compensate for increases by others.
In particular, every country retains the right to increase its resource use in order to reach and maintain basic living standards.

www.overshoot-day.org/

HELPFUL WORDS AND PHRASES
for TASK 3:
My individual ecological footprint ...
- is marked by ...
- could be reduced by ...
- is less than ...
- seems to be ...
- depends on ...

TASKS

1. Describe the term 'Ecological Footprint' in German.
2. Comment on the concept of 'Shrink and Share'.
3. Prepare for a discussion about your personal ecological footprints.

KEY TERMS

- biocapacity
- ecological footprint
- ecological overshoot

City of Leipzig
Recreational lake (former opencast mine)
Protected landscape
Amusement park
Roller coaster
Main entrance
Car park

Sustainable Development

94 Sustainability and Sustainable Development
96 Case Study 1: Great Plains (USA) – Sustainability Through Soil Conservation
98 Case Study 2: Mumbai (India) – Sustainability of Plastic Recycling
100 Case study 3: Indonesia – Sustainable Population Development
102 Think Globally – Act Locally

M1 Sustainable development of a former open cast mining area south of Leipzig (Germany)

Sustainability and Sustainable Development

M1 Sustainable forestry in Germany

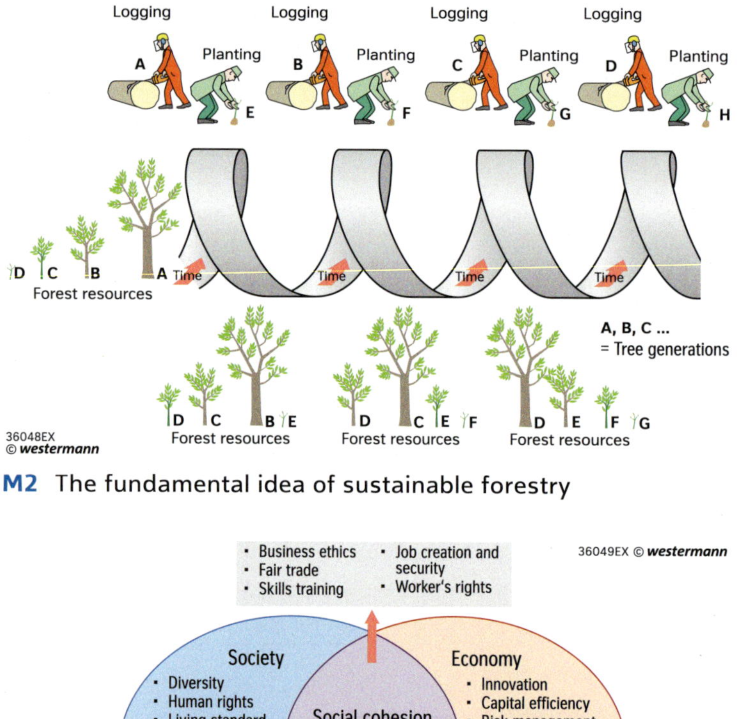

M2 The fundamental idea of sustainable forestry

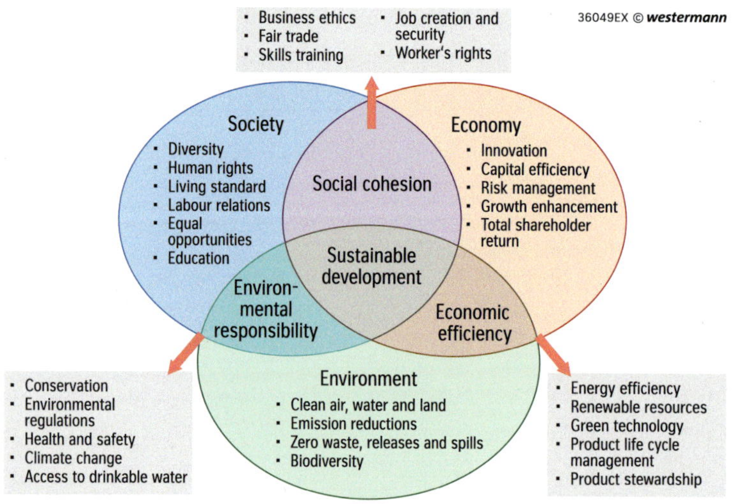

M3 Aspects of sustainable development

The concept of sustainability

In Germany the concept of sustainable planning and managing has been in use in communal forestry since the early 18th century. Due to a crisis in wood supplies for the mining industry at that time, a forest management was demanded which included a systematic afforestation and a lasting (= sustainable) forest use. The aim was to ensure that only as much wood was logged as could grow back since the last logging (**M2**, **M5**).

The concept reappeared with the emerging environmental movement from the 1960s onwards. In 1987, the World Commission on Environment and Development defined sustainable development in the Brundtland Report (**M5**). Since 1992 sustainability was the focus for international politics to create an agenda for the 21st century (›p. 102).

Sustainable development always combines environmental sustainability (ecological aspect), economic growth (economic aspect), and social inclusion (social aspect). The three aspects are linked. They can be used to assess sustainable development. A diagram of interlocking circles is often used to clarify the relationship between the three aspects (**M3**).

The sustainable development of a region is usually a result of manifold long-term efforts. Improving one of these aspects may have positive effects on the other ones (›pp. 96-101).

Sustainable Development Goals

In several conferences, the United Nations' member states have adopted 17 Sustainable Development Goals (SDGs) which should be reached by 2030 (**M4**). There were, in all, 169 proposed targets for these goals and more than 300 indicators to show compliance. The targets have to be implemented on a national level where governments, businesses, NGOs and other stakeholders have to work together. This requires much planning and organisation.

M4 Sustainable Development Goals

1713 *'Wird derhalben die größte Kunst/Wissenschaft/Fleiß und Einrichtung hiesiger Lande darinnen beruhen / wie eine sothane[1] Conservation und Anbau des Holtzes anzustellen / daß es eine continuierliche beständige und nachhaltende Nutzung gebe / weiln es eine unentberliche Sache ist / ohne welche das Land in seinem Esse[2] nicht bleiben mag.'*
([1] solche [2] Bestand)
Source: 'Sylvicultura oeconomica' (H.C. von Carlowitz: Leipzig. 1713)

1974 *'We have not inherited this earth from our parents to do with it what we will. We have borrowed it from our children and we must be careful to use it in their interests as well as our own.'*
M.H. Cass, Australian Minister for the Environment and Conservation, Paris 1974

1987 *'Sustainable development is development that meets the needs of the present without compromising the ability of future generations to meet their own needs.'*
Source: World Commission on Environment and Development: Our Common Future. (Oxford 1987, Frequently referred to as Brundtland Report)

2014 *'We have only one planet. There is no Plan B because there is no Planet B.'*
Ban Ki-Moon, UN General Secretary, New York, 2014

2017 *'Sustainability (ecology): A means of configuring civilization and human activity so that society, its members and its economies are able to meet their needs and express their greatest potential in the present, while preserving biodiversity and natural ecosystems, planning and acting for the ability to maintain these ideals for future generations.'*
Source: Wiktionary, 2017

M5 Sustainability issues

M6 Cartoon

TASKS

1. Explain sustainable forestry with the help of **M2**.
2. Transfer the texts (**M5**) into German.
3. a) Explain **M3** with the help of the latest definition in **M5**.
   b) Analyse the cartoon **M6** with the help of a suitable definition in **M5**.
4. a) Collect further information about one SDG.
   b) Make a poster about one SDG to inform your class.

KEY TERMS

- sustainability
- sustainable development
- Sustainable Development Goals (SDGs)
- sustainable forestry

M1 A duststorm approaches (Kansas, 1935)

"To the (...) part of the (...) country of Oklahoma, the last rains came gently (...). The sun flared down on the growing corn day after day (...) Then it was June, and the sun shone more fiercely. (...) A day went by and the wind increased, steady, unbroken by gusts. The dust from the roads fluffed up and spread out and fell on the weeds beside the fields (...). Now the wind grew strong and hard and it worked at the rain crust in the corn fields. Little by little the sky was darkened by the mixing dust, and the wind felt over the earth, loosened the dust, and carried it away. (...) The dawn came, but no day. In the gray sky a red sun appeared, a dim red circle that gave a little light, like dusk; and as that day advanced, the dusk slipped back toward darkness, (...).

Source: 'The Grapes of Wrath' (John Steinbeck, 1939, Penguin Classics)

M2 The long-lasting darkness

Case Study 1: Great Plains (USA) – Sustainability Through Soil Conservation

Situation

The 150,000-square-mile area in the southern parts of the Great Plains region of the United States has little rainfall, light soil, and high winds. This is a potentially destructive combination. During the 1930s, after years of persistent drought, this region was affected by massive soil erosion. A periode of heavy duststorms called the Dust Bowl turned this fertile farming region into a desperate area. Some geographers also use the term Dust Bowl to define the affected area (**M5**).

Causes

During World War I farmers began replacing native prairie grasses with wheat for military troops. In order to expand the crop land, farmers had ploughed and destroyed the grasses that had kept the stability of the soil. Wheat, a staple crop, does not withstand dry weather conditions, as prairie grass does, and fails to protect land from wind erosion. Additionally, in the 1920s cattlemen overgrazed the region.

When there was the drought in the 1930s, the soil lacked the stronger root system of grass as an anchor, so the winds easily picked up the loose topsoil and swirled it into dense dust clouds.

Consequences

The consequences of the use of improper farming and cultivation techniques concerned various aspects. People suffered from 'dust pneumonia', an illness is caused by dust which fills the lungs. Animals and people, who had this disease, died. Plants and crops were destroyed which lead to a massive loss of harvest. Since fertile soil was blown away, the agricultural productivity was poor. This lead to a rural exodus. By 1940, more than 2.5 million people had fled from regions affected by the Dust Bowl. Ruined farms forced people to leave their homes (**M3**).

Solutions

Nowadays, there is still soil erosion in the Great Plains. However, it is no longer a Dust Bowl. In order to limit soil erosion and to secure this area as a farming region for the future, scientists have developed several measures for soil protection and soil conservation for farmers (**M6**).

M3 Abandoned farm (Texas, 1938)

"And then the dispossessed were drawn west (...) families, tribes, dusted out, tractored out. Car-loads, caravans, homeless and hungry (...). They streamed over the mountains, hungry and restless – restless as ants, scurrying to find work to do – to lift, to push, to pull, to pick, to cut – anything, any burden to bear, for food. The kids are hungry. We got no place to live. Like ants scurrying for work, for food, and most of all for land."

Source: 'The Grapes of Wrath' (John Steinbeck, 1939, Penguin Classics)

M4 On the run

HELPFUL WORDS for M6:

- 1 – konservierende Bodenbearbeitung
- 2 – Mulchsaat

pp. 28-29 Flow Charts

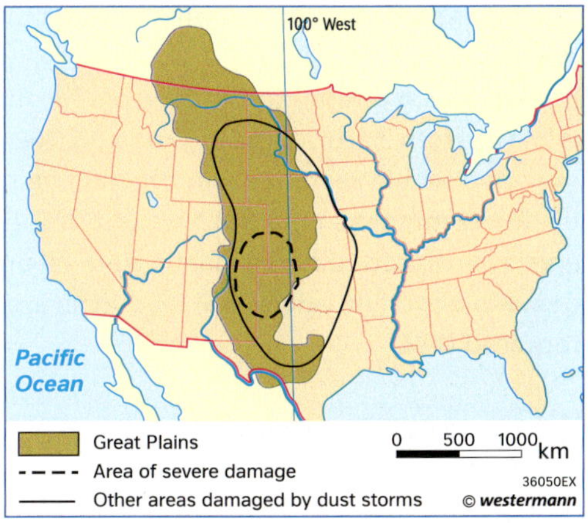

M5 The Dust Bowl

There are several methods of soil protection and conservation that can be achieved through agricultural measures:

- *Minimum tillage*[1]*:* The farmer does not use a plough at the end of the season. Therefore the soil is not turned und there is no loose topsoil which could be blown away by the wind.

- *Stubble-mulching*[2]*:* Stubble is the part of a plant left after the seed has been harvested (e.g. wheat or oats straw). When straw or stubble is kept on the soil surface, the soil is prevented from drying up and from wind erosion.

- *Contour farming:* The farmer ploughs parallel to the slopes. In hilly areas the crops are planted parallel to the slopes.

- *Strip Cropping:* The farmer grows the crops across the prevailing wind direction, not parallel to it. Moreover the crops are grown in strips. Between the row crops like for example soybeans or corn, the farmer plants strips with densely sown crops such wheat or grass. When the wind blows away the soil from crops planted in rows, the soil gets trapped by the cover crop in the neighbouring strip.

- *Windbreaks:* Windbreaks, such as shelter hedges, slow the force of the wind over open ground. Furthermore these shelter hedges support the formation of dew, stabilise the soil moisture and reduce the evaporation. Therefore the soil does not dry out and cannot be blown away so easily.

Contour farming

Strip cropping

M6 Sustainable measures to protect the soil

TASKS

1. a) List the causes of the soil erosion in the Dust Bowl.
   b) Make a flow chart of the causes and effects of the Dust Bowl.
2. Discuss in what way the solutions are sustainable (**M6**).
3. For experts: a) Research on modern ecological challenges on the Great Plains.
   b) Prepare a presentation of your findings.

KEY TERMS

- Dust Bowl
- farming
- soil conservation
- soil erosion
- sustainability

HELPFUL WORDS AND PHRASES for TASK 2:

- Sustainability means ..., so one can say that ...
- First of all, ...
- I would like to point out ...

M1 Plastic on Mumbai's Versova Beach

M4 Ragpickers collecting plastics

Case Study 2: Mumbai (India) – Sustainability of Plastic Recycling

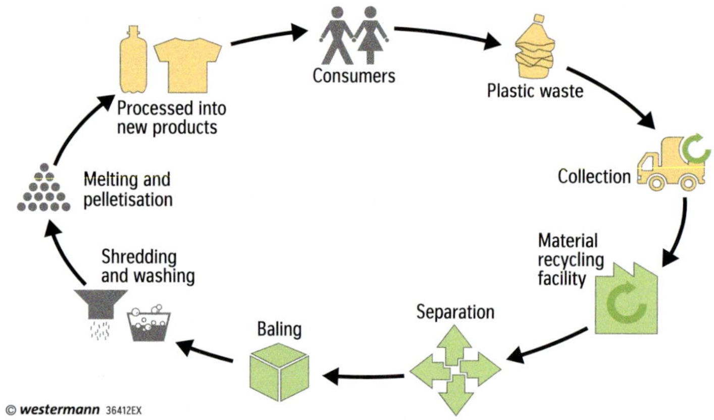

M2 Recycling of plastic in MEDCs

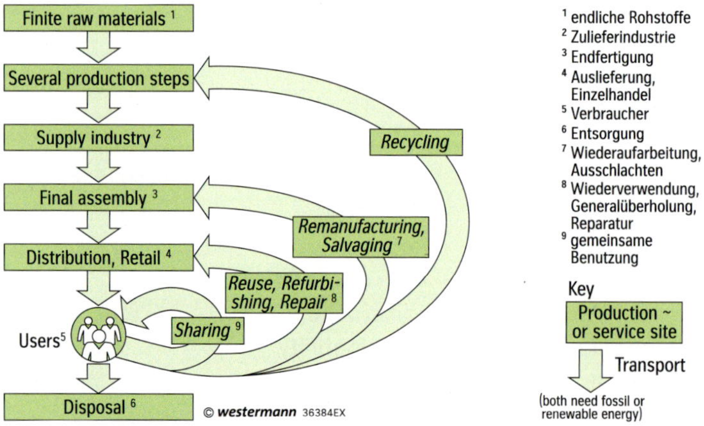

M3 Recycling in the circular economy

Plastic everywhere

Convenient, versatile, light-weight and inexpensive: Around the world plastic is used for many different purposes. However, most of the plastic is used for packaging. It has a short service life and becomes waste after use. Scientists calculated that plastic waste takes more than 400 years to decay.

It takes much energy to produce plastic. It is made from natural gas or crude oil, two non-renewable resources. Recycling plastic, however, uses only ten per cent of the energy required to make plastic from oil or gas. Moreover, no additional resources are required (**M3**).

Waste recycling in LEDs

Urban waste collection is very expensive (**M2**). This is the main reason why it so often fails in less economically developed countries (LEDCs). Waste is often disposed of along roadsides or in rivers, polluting the environment. Most of the solid waste is collected by ragpickers and sold on to dealers. From there, waste finds its way to various processing units where saleable items are made out of it, so they can be reused.

This form of recycling is the most important part of the circular economy in LEDCs (**M3**). It creates jobs, saves resources, protects the environment, and reduces waste on a massive scale.

M5 Sorting plastics in Dharavi

M6 Washing the shredded plastic in Dharavi

Plastic recycling in Mumbai/Dharavi

Dharavi, a district of Mumbai, is the biggest slum in all of India. Supposedly nearly a million people live there on 2 km^2. It became a tourist attraction after the film 'Slumdog Millionaire'.

There is an average of one water toilet for 1,400 people and one water tap per 100 people. The open sewers are the reason for most of the around 4,000 reported cases of diphtheria and typhoid per year. Lung cancer, tuberculosis and asthma are well above the national level.

However, Dharavi is a hub of the recycling industry with up to 15,000 single-room 'factories'. These are mainly part of the informal sector or informal economy.

Most of the work is done manually. Two thirds of the factories are involved in plastic recycling. This way more than 80 per cent of Mumbai's plastic is recycled, a higher percentage than in most MEDCs.

It is said that 200,000 people work in Dharavi's recycling industry. Plastic recyclers are mostly young males, many of them migrant, seasonal workers from other Indian states with little or no education. Although they earn only 3 US-$ per day on average, they at least have employment to contribute to the family income. The workers are often exposed to heavy pollution and much of their earnings is spent on health care.

Informal economy

In LEDCs many enterprises and workers are said to be in the informal economy (informal sector), meaning the working relationships are not formal. They do not show up in official statistics. There are no working contracts, and thus no social security. No taxes are paid by employers and by employees.

pp. 28-29 Flow Charts

pp. 32-33 Texts

pp. 48-49 Tasks

HELPFUL WORDS AND PHRASES

for TASK 2:

- Because plastic is not degradable ...
- Unless there are no other ways to...
- Otherwise especially LED countries ...
- In order to develop a circular economy for plastic ...

for TASK 3:

- In order to increase plastic recycling ...
- I suggest ...
- On one hand there is the advantage of ..., on the other hand we have to consider ...;
- To reduce the considerable health risks ...

TASKS

1. Describe the necessity of plastic recycling (Text, **M1**).
2. Explain in what way the recycling of plastic is sustainable (Text, **M2**, **M3**).
3. Discuss ways to increase sustainable development of plastic recycling in Mumbai.
4. For experts: Contrast the plastic recycling industry in Dharavi with that in your home town (in German).

KEY TERMS

- circular economy
- informal economy
- informal sector
- ragpicker
- recycling

M1 'Come on, two kids are enough' – street signs are part of a long running family planning campaign in Indonesia

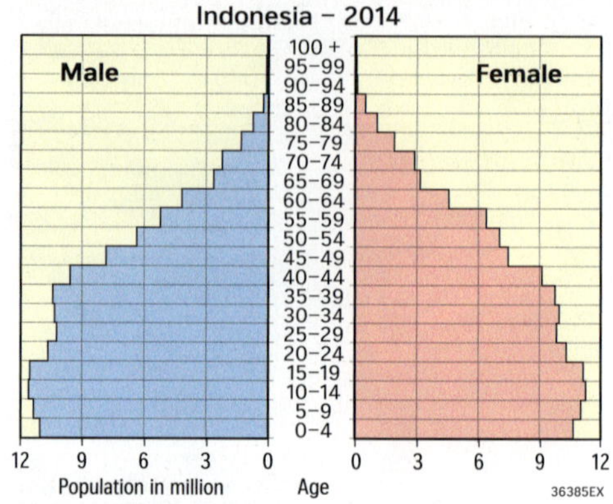

M3 Indonesia – age-and-gender diagram

Case study 3: Indonesia – Sustainable Population Development

'My solution is exploit more land – because if you exploit all the land in Indonesia you can feed 250 million, and I have only 103 million [...] in my country, the more (children), the better.'

Sukarno, first President of Indonesia (1945-1967)

Rising pressure

The total fertility rate in Indonesia reached its peak during the 1950s. President Sukarno was of the opinion that a large population meant national wealth and strength.

However, a fast rising number of people led to worse living conditions for many Indonesians. More people meant more pressure on scarce agricultural resources, especially on Indonesia's most densely populated island of Java.

Such population growth is of course not sustainable. The people needed more space for farming, and slash-and-burn agriculture started to have a huge impact on the natural environment, mostly tropical rainforests. In addition, the poor soils could not provide food for a long time, and new farmland had to be cleared for usage.

Also, a shrinking area of arable land per person led to migration from rural areas into the cities, forming a constantly growing group of the urban poor.

Family planning

Supply shortfalls in rural areas, hunger, poverty, and rising criminality in the fast growing cities were reasons to act politically (**M1**). During the 1970s, the Indonesian family planning programme Keluarga Berencana was introduced to slow down population growth by reducing the total fertility rate of 5.6 children per woman.

Family planning was advertised all around the country. Mobile family planning teams including nurses and midwives educated

100800-196
schueler.diercke.de

DE-114
www.diercke.com

pp. 16-17
Tables

pp. 152-153
Age-and-Gender Diagram

Year	Total population (million)	Growth rate (%)	Fertility rate (children per woman)	Agricultural area (ha per person)	Urban population (%)
1965	100.3	2.70	5.6	0.179	15.8
1975	130.7	2.52	5.0	0.138	19.3
1985	165.0	2.12	3.7	0.118	26.1
1995	197.0	1.54	2.7	0.088	36.1
2005	226.3	1.33	2.5	0.101	46.0
2015	257.6	1.21	2.4	0.092*	53.7

Source: Worldbank (2017), *2014

M2 Population development of Indonesia (1965-2015)

Indicator	Past	2015
Life expectancy (years)	51.7 (1965)	69.1
Adult literacy rate (%)	67.3 (1980)	95.4
GDP per capita (PPP in US-$)	2,989 (1990)	11,058
Poverty rate (%)	70.3 (1985)	10.9
Forest area w(% of land area)	65.5 (1990)	50.2
		Source: Worldbank (2016)

M4 Indonesia – selected indicators

M5 Smoke caused by slash-and-burn in Sumatra

the rural population about matters of family planning. They distributed modern contraceptives within the communities and took care of health issues mainly of mothers and their children.

In a final phase, the programme aimed at establishing family planning on a larger scale in the communities, and hence in the society as a whole, to ensure a lasting future success.

Success, but still problems

The family planning programme is widely regarded as successful (**M2**). However, due to the longterm impacts of demographic transition, the Indonesian population is still growing (**M3**).

Illegal slash-and-burn still contributes to the destruction of rainforests. Since most of the young people look for jobs in urban areas, population density problems due to urbanisation are further increasing (›p. 53).

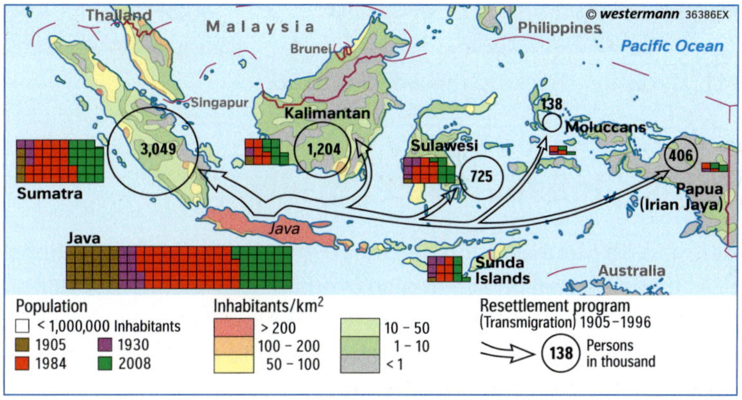

M6 Indonesia – population density/population development

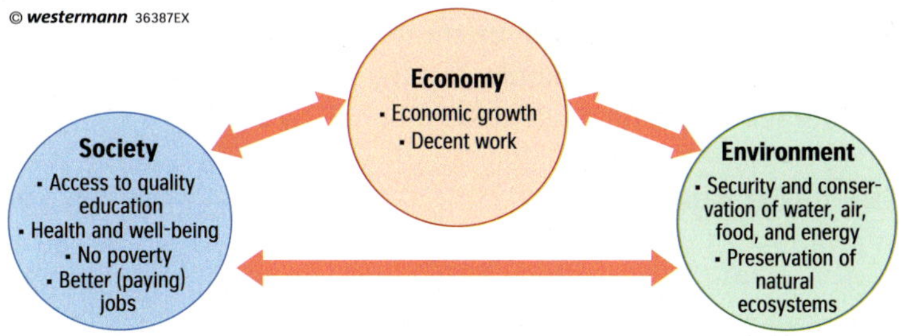

M7 Indonesia's sustainable development goals

TASKS

1. Name social, environmental and economic problems of Indonesia in 1965 mentioned in the text.
2. Explain Indonesia's population development between 1965 and today (Text, **M2**, **M6**).
3. Discuss ways in which family planning can help Indonesia reach its sustainable development goals (**M7**).

KEY TERMS

- demographic transition
- population growth
- sustainable development
- urbanisation

M1 Cartoon – 'Kicking the world around'

ICLEI – Local Governments for Sustainability

ICLEI is the most important global network of more than 1,500 communities committed to building a sustainable future. The network includes cities, towns and regions from all around the world and represents over 25 per cent of the global urban population. By developing green economies and smart infrastructures, ICLEI-members want to become low-carbon, ecomobile, more resource-efficient, biodiverse, productive, healthy and happy as well.

Think Globally – Act Locally

At the Earth Summit in Rio de Janeiro in 1992, the United Nations (UN) implemented an action plan (Agenda 21) in which sustainable development was recognised as the major challenge. The targets of Agenda 21 aim at giving mankind the opportunity to overcome the challenges of the 21st century (**M1**).

For sustainability, a change is necessary in almost every area of life in all levels of society, from the global to the local perspective. In 2012, at the anniversary conference Rio+20, the attending members reaffirmed their commitment to Agenda 21. A statement of intent to promote sustainable development was drawn up and it was agreed to define goals to be reached by 2030 (›p. 94).

The Local Agenda 21

In Chapter 28 of Agenda 21 all governments are requested to encourage their citizens personal sustainable actions. To this end, beginning in the middle of the 1990s, many communities around the world launched Local Agenda 21 (LA21).

Priority targets in LEDCs are health care, fresh water supply, development in rural areas, and fighting poverty. In MEDCs targets of the LA21 mainly concentrate on sustainable economic policies dealing with energy consumption, agriculture, and trade.

The LA21 should always aim at initiating a discussion process in which the local council, the administration, the advisory board as well as churches, political parties, trade unions, research institutes, and citizens are involved. The communities are obliged to develop programmes, and to set up projects that are environmentally sound, socially agreeable as well as economically justified.

IntegraBike: Pedaling towards sustainable urban development

With over 100 km of bicycle paths, Sorocaba is a national reference point for excellence in urban mobility. Its successful free bike sharing program, IntegraBike, not only promotes the integration between transport modes, but has also improved the quality of life among Sorocaba's citizens, and contributes to the reduction of greenhouse gas emissions. [...]

Source: www.iclei.org,
ICLEI Case Study – No.184, 2016, Sorocaba, Brazil

M2 ICLEI case study (Excerpt)

More ICLEI case-studies: www.iclei.org/activities/resources/publications

M3 Bicycle path in Sorocaba/Brazil

M5 Free vegetables in a public garden in Andernach

Start sustainability at home!

There are many starting points for your personally sustainable behaviour.
Do not forget that electricity consumption through standby wastes 20 million kilowatt hours of electricity in Germany per year. By way of comparison: The city of Berlin consumes 14 million kilowatt hours in the course of a year.
Also you can reduce the amount of energy you use for heating and hot water. And there is much more you can do than saving energy (**M4**).

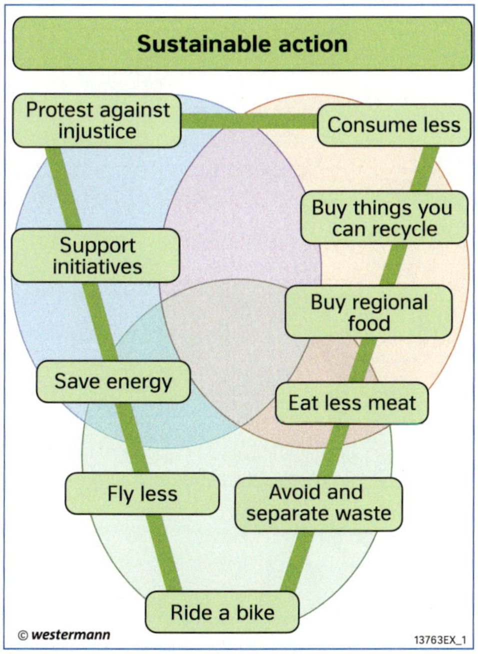

M4 Things that can be done

Local Agendas – German Examples

Urban Gardening / Andernach

This project is one of the biggest urban gardening projects in Germany. Citizens are allowed to pick and harvest vegetables in public vegetable patches in the middle of the town. Andernach proudly presents itself as the first 'eatable' town in Germany.

Reparaturcafé / Düsseldorf

A repair café is a meeting/event where broken gadgets can be repaired. Instead of throwing the things away, they are fixed with the help of voluntary experts in a friendly and lively atmosphere.

Energy-adventure-path / Nürnberg

This energy-adventure-path for kids is an example for sustainable education. The mobile exhibition explains aspects about the energy production and consumption in Germany in a vivid and practical way.

TASKS

1. List goals of the Local Agenda 21.
2. Explain the sustainability behind the examples shown on this page.
3. For experts: a) Research on further LA21 projects of ICLEI-members.
   b) Make a poster about this project.

pp. 154-155 Cartoons

HELPFUL WORDS AND PHRASES
for TASK 2:
- By growing vegetables in ...
- A positive consequence / result / ... could be ...
- By using ... you support ...
- The usage of ... can be sustainable, because it helps ...
- To learn / know about ... is important, because ...

KEY TERMS

- Agenda 21
- Local Agenda 21
- sustainability

Flag of the EU
National flag of Croatia
Newspaper
24 SATA
EUROPA!
Headline
ERSTE

Europe Changes

106 The European Union (EU) and Me
108 The European Union and the World
110 Regional Disparities in the EU
112 The Regional Policy of the EU
114 The Common Agricultural Policy of the EU
116 Connecting Europe – Case Examples
118 The European Car Industry Moves East
120 The EU – Quo Vadis?

M 1 Young people celebrating the entry of Croatia into the European Union (2013)

The European Union (EU) and Me

M1 The European Union member states

100800-084 schueler.diercke.de

DE-039 www.diercke.com

A The EU supports common educational goals and programmes. Standards were developed to categorise language skills (B2, C1, etc.). Furthermore, your school reports as well as training certificates are accepted throughout the EU. This makes it easier for you to study in another European country or to do voluntary service there.

B EU regulations protect customers, which is why telephone and internet costs have been reduced over the past years. In June 2017 roaming charges were entirely abolished.

C Your school exchange programmes are quite likely also supported by the EU, as financial help is provided, for instance, by the Comenius programme.

D Thanks to the European Health Insurance Card (EHIC), which is accepted all throughout the EU, visits to a doctor and even treatments at hospitals are simplified as you are guaranteed state-provided healthcare.

E Due to the Schengen Agreement, you are free to move in the EU and do not even have to show your passport at the border.

F Travellers have been given more rights with respect to delays or cancellations over recent years. This is not only true for train passengers, but also for those travelling by plane, ship or bus.

G The common currency, the Euro, makes life rather convenient in 19 of the 28 member states of the EU. It allows you to travel without having to change money and to compare prices easily.

H Owing to the cooperation of formerly independent countries, the EU ensures safety for its citizens. Therefore, you now live in relative peace and stability.

HELPFUL WORDS AND PHRASES

for TASK 2:

- My first choice is … as it is very important to me that …
- Furthermore / Moreover I think …
- Consequently, …

TASKS

1 Categorize of the statements (A–H) about life in the EU and find suitable headings for each category.

2 a) Sort the statements (A–H) according to their importance to you.

b) Give reasons for your arrangement.

c) Compare your results.

KEY TERMS

- Euro
- European Union
- Schengen Agreement

M1 Examples for different living conditions in Europe – Vulcan (Romania) and Amsterdam (Netherlands)

Regional Disparities in the EU

100800-099, -101
schueler.diercke.de

DE-040
www.diercke.com

pp. 10 - 11
Thematic Maps

HELPFUL WORDS AND PHRASES

for TASK 3:

- One advantage / disadvantage is that ...
- Consequently ...
- One can see that ...

There are great regional disparities within the EU. These differences are denoted by an unbalanced distribution of e.g. natural resources, industrial sites, as well as cities. Consequently, the disparities also result in differences concerning the employment structure, living standards, and the concentration of economic activities.

Taking a closer look at the EU member states, one can see that they have a significantly divergent economic power. With each enlargement of the EU in 2004, 2007 and 2013, the polarization within Europe has steadily increased.

Wide gaps exist between the former socialist countries in the east e. g. Hungary, Romania or Poland, and the western European states.

Furthermore, there are also great imbalances within the single states: economically underdeveloped regions, so-called deprived areas, are right next to economically strong ones. For example, in France or Germany there are unprogressive rural areas with few jobs and much improvable infrastructure, as well as agglomerations, shaped by the tertiary sector.

	Life expectancy at birth	Population with at least some secondary education (in %, ages 25 and older)	Physicians per 10,000 people (2001 – 2013)	People at risk of poverty or social exclusion (in %)
Bulgaria	74.2	94.3	38.1	41.3
Denmark	80.2	96.1	34.2	17.7
France	82.2	80.5	31.8	17.7
Germany	80.9	96.6	38.1	20.0
Greece	80.9	63.2	43.8	35.7
Hungary	75.2	98.3	29.6	28.2
Latvia	74.2	98.9	28.8	30.9
Poland	77.4	82.3	22.0	23.4
Romania	74.7	88.9	23.9	37.4

http://ec.europa.eu/eurostat/de/web/sdi/indicators
http://hdr.undp.org/sites/default/files/hdr_2015_statistical_annex.pdf

M2 Statistical data (2015)

The Blue Banana

The basic model of the so-called 'Blue Banana' was developed by Roger Brunet in 1989 to make aware of the shortcomings of the French regional (planning) policy. Today, this model is the most used one to describe the current state and the development potentials of European regions. It was also used as a basis for further models.
Within this belt, which covers an area 1,300 x 900 km^2, are very important production sites. It is an economic corridor that also fostered transnational collaboration. The basis for this collaboration was, however, created through the establishment of the European single market.

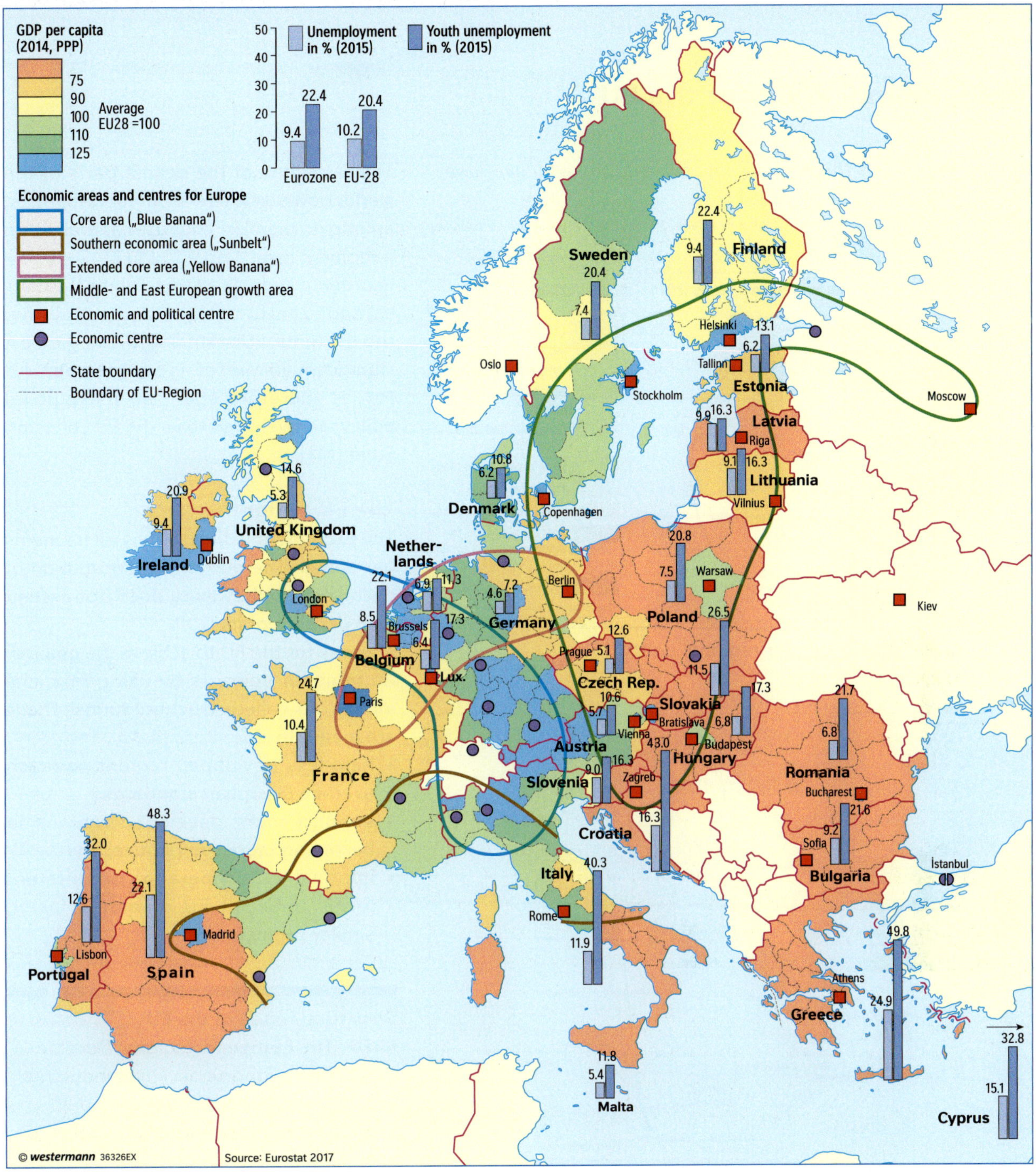

M3 Regional disparities in Europe

TASKS

1. Describe regional disparities within the EU.
2. Compare the data provided in **M2**.
3. Contrast the regions covered by the 'Blue Banana' to the neighbouring regions.
4. For experts: Discuss in how far it is possible to reduce or even compensate disparities within Europe. Use your atlas.

KEY TERMS

- agglomeration
- living standard
- regional disparitiy
- rural area
- tertiary sector

The Regional Policy of the EU

'Cohesion policy is the main EU investment tool, with over one-third of the Union budget.' (2013)

Johannes Hahn, Member of the European Commission

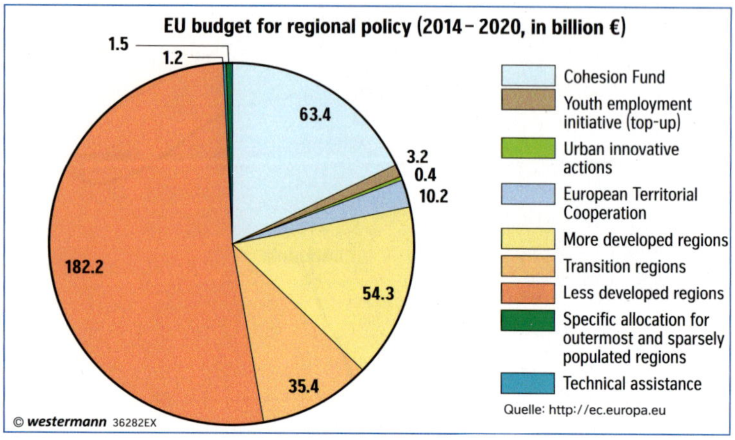

M1 EU budget available for regional policy

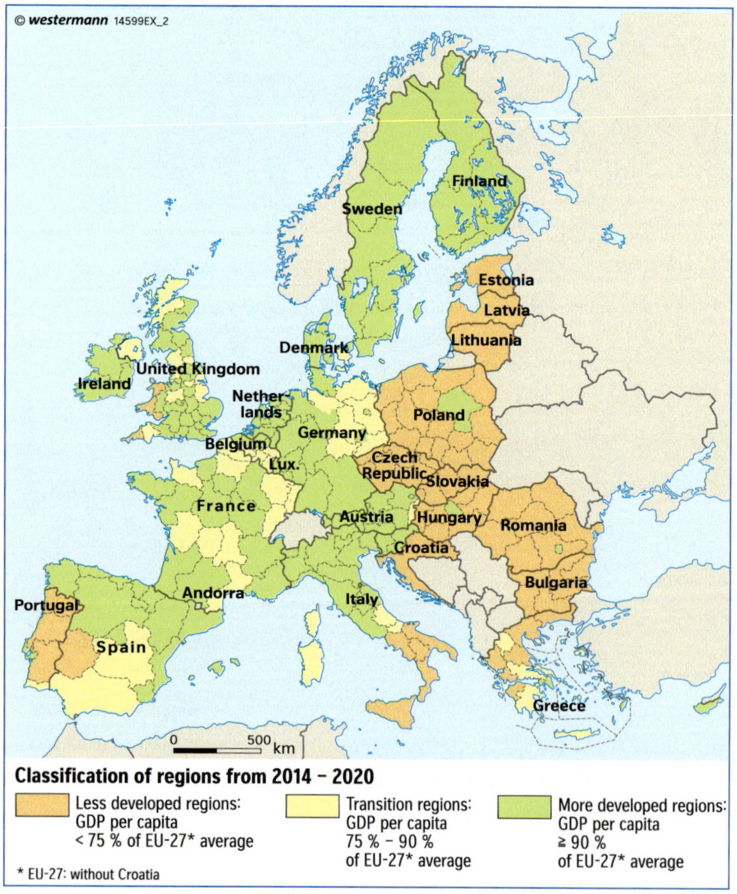

M2 EU-classification for regional policy (2014 – 2020)

The EU is one of the richest parts of the world. However, there are great disparities, among its member countries as well as on a regional level (› pp. 110 / 111).

In order to support a strong domestic market and to ensure a high standard of living, the EU aims at balancing these gaps. A significant amount of its budget (› pp. 112: **M1**) is therefore spent on its regional policy, also referred to as the EU Cohesion Fund (**M1**).

The EU budget

The EU budget is financed by all its member states. The maximum contribution of each country, however, does not exceed 1.23 per cent of its GDP.

The way for the EU to achieve the goals of its regional policy is by using financial means, so-called structural funds. These are meant to

- help less developed regions to reach average European standards,
- support more developed regions in becoming more competitive, and
- improve the cooperation on different levels between EU members facing similar problems.

Structural funds of the EU

ERDF: The European Regional Development Fund supports declining industrial regions.

ESF: The European Social Fund helps to ensure jobs.

CF: The Cohesion Fund concentrates on improvements concerning infrastructure and the environment.

EAFRD: The European Agricultural Fund for Rural Development focuses on solving problems in agriculture.

EMFF: The European Maritime and Fisheries Fund manages fishing fleets and conserves fishing stocks.

Saxony-Anhalt – a transition region

In the course of the reunification of Germany in 1990, all parts of East-Germany became convergence regions within the EU. Although these states have already undergone a major process to catch up with the other German states, they are still lagging behind.

With financial support from the European Regional Development Fund (ERDF), totalling 1.4 billion € between 2014 and 2020, the German state of Saxony-Anhalt is striving to build a future-oriented, competitive, and thus sustainable economy. The focus is on the following five goals:

- stimulating research, development and innovation,
- supporting small and medium-sized enterprises (SMEs) and thus improving their competitiveness,
- reducing CO_2-emissions,
- climate protection and risk prevention,
- and supporting urban development.

Möller Industrietechnik GmbH

The company is located in Hecklingen, Saxony-Anhalt. It is the third biggest full-range trader of seals in Germany. Various types of industries use their products including the pharmaceutical, automobile, food and chemical industries.

Ever since its founding in 1994 with a start-up capital of 30,000 DM, the company has steadily been expanding. This process has been supported by eight government grants which were used to increase the product range, install new production lines, and successively raise storage capacity (M3, M4). Most recently it received over 400,000 € from the ERDF.

M3 Müller Industrietechnik in the industrial estate in Hecklingen (1998 and 2011)

Raum für Wachstum

'Ohne die Förderung durch die Europäische Union hätten wir unser Wachstum so nicht realisieren können [...]. Mit Hilfe der Investition in den Ausbau der Lagerkapazitäten konnten wir die Erweiterung der Sortimentspalette logistisch absichern, vor allem die Belieferung der Händler zu beschleunigen. Das schafft und sichert Arbeitsplätze', betont Ralf Möller.

Mit rund 150 Beschäftigten und sieben Auszubildenden ist die Möller Industrietechnik GmbH sowie die 2009 gegründete Möller Metalldichtungen GmbH nicht nur der drittgrößte, sondern auch einer der attraktivsten Arbeitgeber im Salzlandkreis [...].

'Unsere Leute stammen alle aus der Region. Denen wollen wir hier vor Ort eine sichere berufliche Perspektive geben, von den jungen Leuten bis hin zu den älteren Arbeitnehmern' [...].

Source: www.europa.sachsen-anhalt.de (Kai Bieler, Apr 2017)

M4 An example from Saxony-Anhalt

http://ec.europa.eu/regional_policy/en/atlas/programmes

HELPFUL WORDS AND PHRASES

for TASK 1:
- With new members joining ...
- Disparities between ...
- In order to balance ...

for TASK 2:
- The most money is given to ... because ...
- About ... million Euro are granted to ... as ...

for TASK 4:
- One programme is ...
- Its major goal is to ...

TASKS

1. Describe the regional disparities within the EU (M2).
2. Explain why structural funds are a suitable measure to compensate disparities.
3. Explain the importance of Möller Industrietechnik GmbH for the region (M4).
4. For experts: a) Research on another project in your home state, financed by the EU. b) Prepare a presentation about it.

KEY TERMS

- regional disparitiy
- regional policy
- structural funds

Connecting Europe – Case Examples

Fast interconnections are essential for a successful European economic zone. Therefore the European Union was entrusted in the Treaty of Maastricht with the task of constructing and expanding Trans-European Networks (TEN) in the fields of transport, communication and energy. These networks are supposed to contribute to the further development of the European single market and thus to strengthen the economic and social cohesion of the EU member states.

The latest infrastructural policy plans to transform the current patchwork of streets, rails, airports and canals across Europe into a single, homogeneous trans-European network (**M1**). The goal is to guarantee regional and national transport networks for humans and goods, which are accessible and affordable for everyone.

M1 TEN transportation core corridors

M2 Planned 'Main line for Europe'

M3 TGV and ICE

'Main line for Europe'

One project in this regard is the 'Main line for Europe' (TEN-T / Project 17), a railway passage connecting Paris to Budapest – also including a link to the city of Bratislava, Slovakia (**M2**). Being located half way between the North Sea and the Mediterranean, this is a vital east-west axis in Europe, which will facilitate economic, political and cultural contact and exchanges between Western and Eastern Europe.

In the light of 35 million people living in the commuting area of the railway in France, Germany, Austria, Slovakia and Hungary, and 16 million employees residing there, the significance of this project is reinforced. Each country has a unique contribution to make to the rail link. France, for example, built the LGV Est européenne, a high-speed link between Paris and southern Germany. The route is not only used by the French TGV, but also by the German ICE (**M3**), both of which reach a speed of up to 320 km/h. As a result of these new direct connections, the travel time between Paris and Vienna will be reduced from 13.5 hours to 9.5 hours.

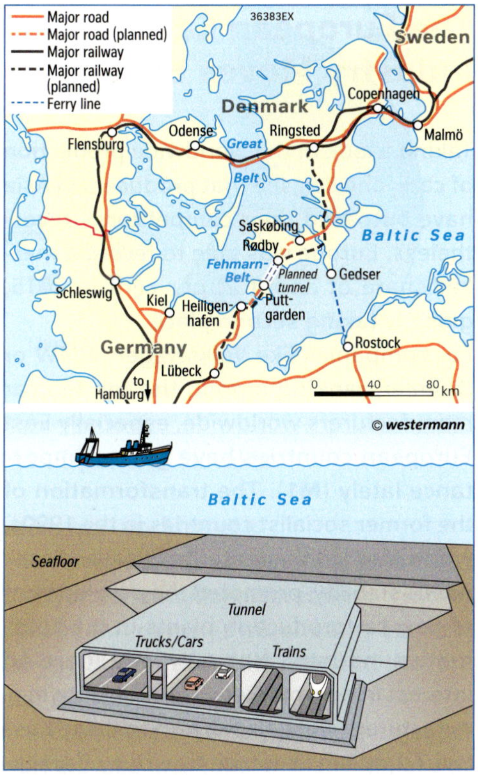

M4 Planned Fehmarn Belt Fixed Link

'This is going to be an economic fiasco! Taxpayers have never been told the whole truth by politicians! Toll revenues will never ever be enough to repay the loans, secured by government guarantees, to the state-owned Femern A/S within 39 years. The traffic forecasts underlying this promise are totally unrealistic ...'

Danish traffic expert
Source: www.shz.de („Dänen warnen vor dem Tunnel-Fiasko", Frank Jung, 14 May 2015)

'The Fehmarn Belt is going to create employment, apprenticeships as well as a strong traffic corridor to Germany. It is also going to shorten journey times for Danish commuters between the capital region and Lolland-Falster.'

Danish politician
Source: http://femern.com/de (18 Mar 2016)

Auf der idyllischen Insel Fehmarn haben viele Bürger Angst davor, dass die Hochgeschwindigkeits-Bahnstrecke und die dann vierspurig ausgebaute Bundesstraße [...] ihre Insel zerschneidet. Sorgen bereiten auch mögliche Rückgänge bei den Touristenzahlen, der Verlust von Agrarflächen und negative Einflüsse auf die Meeresumwelt.
'Vor allem aber stehen die Kosten des Bauwerks in keinem Verhältnis zum Nutzen und dem verkehrlichen Bedarf.'

Aktionsbündniss gegen eine feste Fehmarnbeltquerung
Source: www.welt.de (Eva-Maria Mester, 21 Dec 2015)

M5 Comments on the planned Fehmarn Belt Fixed Link

The Fehmarn Belt Fixed Link

The crossing of the Fehmarnbelt is another project in this context (TEN-T / Project 20), connecting the German island of Fehmarn with the Danish island Lolland. The goal is to guarantee an even better connection between the metropolitan areas of Hamburg and Copenhagen (**M4**).

Although it has long been argued whether a bridge might be the better solution, it is now decided that the Fehmarn Belt Fixed Link below the Baltic Sea will be – with a length of 17.6 km – the longest immersed tunnel. Even though the preparations for the actual construction began in 2013, the completion of the entire project (approved by the Danish government on April 28, 2015) is momentarily scheduled for 2024 at the earliest.

Upon completion, it will be possible for cars to pass the entire distance between Germany and Denmark within ten minutes – in seven for trains.

100800-104
schueler.diercke.de

DE-054
www.diercke.com

TASKS

1. a) Describe the structure of the TEN-T transportation core corridors (**M1**).
   b) Compare the corridors with maps on the subjects of European economy and traffic. Use your atlas.
2. a) Locate the Fehmarn Belt (**M4**).
   b) Describe the planned construction using **M4**.
3. Research on the current status of the Fehmarn Belt Fixed Link.
4. Explain why projects like the ones presented cut both ways (**M5**).

HELPFUL WORDS AND PHRASES
for TASK 2:
- The tunnel connects ...
- Four separate tubes ...
- It provides space for ...

KEY TERMS

- infrastructure
- metropolitan area

The EU – Quo Vadis?

'The European Union is probably the most important ever project for peace. This started after the war - putting together former enemies. It started with six countries and we are now 27, another one is going to join us next year and more want to come. So the EU is the most important project for peace in terms of transnational, supernational co-operation.'
(José Manuel Barroso, President of the European Commission 2004-2014) Source: www.bbc.com (12 Oct 2012)

M1 Statement about the importance of the EU

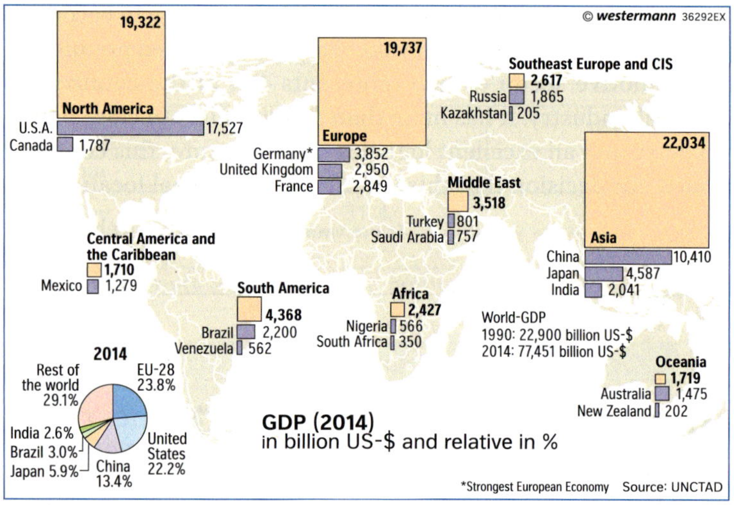

M2 Economic importance of regions

Country	Satisfaction with EU-democracy (in %)	Confidence in the EU-		
		Parliament (in %)	Commission (in %)	Central Bank (in %)
Denmark	68	63	60	61
Germany	48	47	39	32
Greece	28	32	22	19
Poland	61	46	46	37
Romania	64	63	62	52
Spain	38	30	27	22
United Kingdom	40	29	27	26

Source: www.bpb.de (2016)

M3 Attitudes towards the EU (selected countries, 2015)

M4 The EU between celebrations and demonstrations (London)

Achievements

The European Union is an institution that has existed, including all its stages of development, for about 60 years now. It has always advocated the unity of independent countries and thus succeeded in turning former enemies into cooperating partners. As the Norwegian Nobel Committee stated in awarding the Nobel Peace Prize back in 2012 (**M4**): 'The stabilizing part played by the EU has helped to transform most of Europe from a continent of war to a continent of peace.'
Economic prosperity also ranks amongst the goals of the EU. Numbering 28 member states in 2017, it forms the biggest common market in the world.
The creation of this market did not only stimulate trading as well as economic activity, it also raised the standard of living. Four basic freedoms are guaranteed as market functions:

- free movement of goods,
- free movement of people,
- free movement of services,
- free movement of capital.

However, how satisfied people are with a political system also depends on their impression of whether or not their vote counts, and if there is a corresponding influence on political decisions (**M3**).

Challenges

There are, however, some countries which are less content with the EU than others. Additionally, Euro-skepticism has been growing lately (**M3**).
Although the foundation of the European Union let some citizens reconsider their identity, eventually resulting in a sense of European belongingness, recent developments show that national consciousness still dominates.

One of the predominant fears on a national level is the loss of self-determination. While regulations were introduced by the EU to establish a common market, the individual states had to partially give up their national sovereignty – with some perceiving more and more loss over time.
As a British politician put it in 2016:
'We are seeing a slow and invisible process of legal colonisation, as the EU infiltrates just about every area of public policy.'

Thus, individual states also make varying demands on the EU:
• *Fiscal policy:* mainly Southern European states ask for additional exceptions to the European debt policy, while net contributors (**M6**) argue that the Eurozone and the EU can only be successful if debt and stability criteria are satisfied.
• *European borders:* within the scope of the refugee crisis since 2015, one thing has become obvious: country location within Europe matters as geographic location dictates differences in challenges and the related expectations held in an EU member state. Currently many Europeans fear poverty immigration and foreign infiltration by other cultures and religions, especially by the Islam.
• *Individual state supremacy:* in a confederation of states, all members assume equal rights. Nevertheless, some states complain about a German supremacy. Questions like 'Who made the decision?', 'Who is in charge?' or 'Whose word counts?' currently matter more than in the past.

On 23rd June 2016 Great Britain held a referendum in which the country decided on the so-called *'Brexit'* (a merger of the words *Britain* and *exit*).
With a voter turnout of 71.8 %, the British population decided with 51.9 % of the vote to leave the European Union.

Source: Picture from the cover of 'The Spectator', (18 Jun 2016)

M5 The Brexit

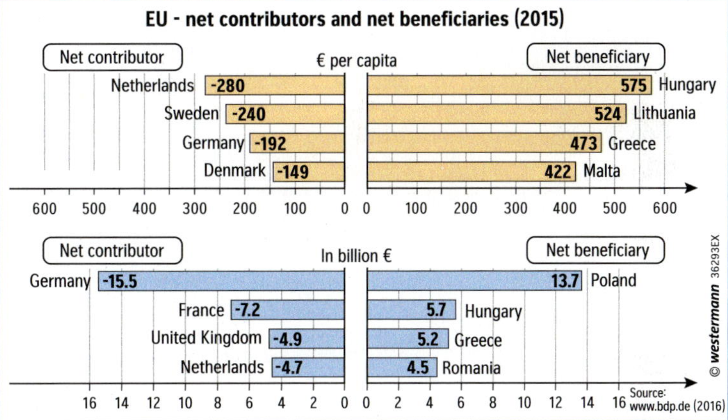

M6 Net contributors and recipients

TASKS

1. List the achievements of the EU (**M1**, text).
2. Explain why more and more people become dissatisfied with the EU.
3. Analyse the cartoon (**M5**).
4. Research on scenarios about the possible future of the EU.

pp. 154 - 155 Cartoons

HELPFUL WORDS AND PHRASES

for TASK 1:
• One success / triumph / accomplishment of ...
• Without the existence of the EU ...

for TASK 2:
• to feel boxed in ...
• to emerge from ...

KEY TERMS

• Brexit
• Euro-skepticism
• European Union

National flag of Germany
Visitors
Info display

Germany in a Changing World

124 Germany – A Leading World Economy
126 Demographic Change
128 A Land of Contrasts – Regional Disparities in Germany
130 Regional Migration in Germany
132 Migration to and from Germany
134 A Bright Future? – Germany's Energy Transition
136 Structural Changes in Agriculture
138 Case Study: Meat Production South of Oldenburg
140 Rise and Fall of the German Coal Industry
142 Structural Change in the Ruhr Area
144 Metropolitan Region Rhine-Main
146 Munich – A Success Story
148 Shrinkage and Growth in the Halle-Leipzig Region
150 Berlin – The Fast Changing Capital

M1 Inside the Reichstag, Berlin, home of the German Parliament

M1 Germany is worldwide renowned for manufacturing and engineering

Germany – A Leading World Economy

Today, Germany has not only become the most powerful economy in Europe, it has also developed into one of the largest national economies in the world.

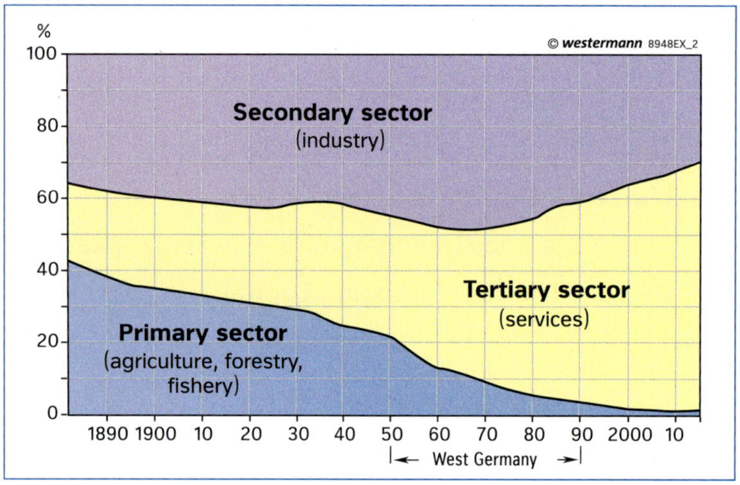

M2 People employed in economic sectors

FACT SHEET • German Economy

- **World ranking:** 5 (GDP, PPP), 3 (Export) 3 (Import), 5/2 (Direct foreign investments at home/abroad)
- **Contribution to Germany's GDP:** Primary sector: 1 %; Secondary sector: 30 %; Tertiary sector: 69 %
- **Total value of export goods:** 1195 billion Euros
- Share of exports to EU member states: 59 %
- **Total value of import goods:** 948 billion Euros
- Share of imports from EU member states: 56 %
- **Balance of trade:** 217 billion Euros

M3 Basic information about Germany's economy (2015)

Changes in the economy

Germany's leading position is the result of a development, which has been going on for more than a hundred years. At the end of the 19th century, most people worked in agriculture. During the age of industrialisation the need for workers was very high as production was labour intensive. Today, due to automation, mechanisation and computerisation, agricultural and industrial production have become less labour intensive. Germany's industrial production concentrates mainly on high technology products. As further research and development is needed, scientists and engineers, who work in the tertiary sector, are also involved in production.
Germany's reputation as a leading economy is mainly based on the export of high quality goods. The slogan 'Made in Germany' has become well known all over the world.

Economic sectors

Primary sector: The extraction of raw materials from the ground and the sea (e. g. agriculture, forestry, fishery).
Secondary sector: Processing of raw materials and manufactured goods (e. g. car industry).
Tertiary sector: Provision of services in trade, finance etc. Industries providing information services, such as computing, ICT, consultancy, and research can be referred to as Quarternary sector.

M4 Main export partners

M5 Main export goods

Strengths

Education, skilled workforce, innovation and research, entrepreneurship, facilities and infrastructure, balanced economy incl. industry and services

Weaknesses

Demographic challenges, lack of investments in new infrastructure, excessive regulation (bureaucracy)

M6 Assessing the German economy

The UK could learn a lot from Germany's long-term industrial strategy

As a share of its economy, Germany's manufacturing sector is twice the size of Britain's [...]. Germany [...] has more world-class companies in a broader range of sectors – Siemens, BMW, Bosch, to name a few. [...]
Germany's industrial power was built on a core of family-owned businesses. [...] Whereas UK companies were often hindered by an overvalued pound, the mark was undervalued, making German exports extremely competitive[1] in world markets.
Another part of the story involves the structure of German industry. The emphasis[2] on vocational education[3] combining academic studies and on-the-job training for apprentices[4] is globally admired. [...]
The final piece of the tale involves the support given by the state. [...] The German government has also used tax incentives to foster industries seen as having growth potential – the development of the solar panel sector through feed-in tariffs[5] being a recent example. [...]

Source: www.theguardian.com
(Larry Elliot, 30 Mar 2016, abridged)

M7 Germany – a model?

TASKS

1 a) Describe the economic development of Germany.
b) Explain its effect on the economic sectors.

2 Give reasons for Germany's present position in the world economy.

3 For experts: Prepare a speech about 'Germany as a leading world economy' in German or English.
Use all the key terms.

100800-070 schueler.diercke.de

DE-041-03 www.diercke.com

pp. 20 - 21 Pie Charts

pp. 26 - 27 Bar Charts

HELPFUL WORDS
for M7:
1 – wettbewerbsfähig
2 – Schwerpunkt
3 – Berufsbildung
4 – Lehrlinge
5 – Einspeisetarif

KEY TERMS

- balance of trade
- economic sector
- high-quality good
- import / export good
- national economy
- primary / secondary / tertiary sector
- world economy

M1 Cartoon – Germany's future population structure

Demographic Change

Germany's population structure has always been changing in terms of age, sex and life expectancy. For over 40 years, the birth rate has been lower than the death rate, making Germany's negative rate of natural increase (› pp. 42 / 43) is one of the lowest in the world. However, this has not led to a rapid decline in Germany's population yet. Due to the immigration in the last 45 years (**M3**), the population was still increasing slowly until the year 2002 (82.5 million inhabitants). But since then, there has been a slight decrease (2015: 81.1 million inhabitants). If the rate of natural increase remains the same, Germany's population will fall to only 25 million in the year 2100 without more immigration.

Problems

The age structure of the population across Germany varies immensely. Amongst other things, this is a consequence of internal migration (› pp. 130 / 131).

In Eastern Germany as well as in the Ruhr area, for example, the share of elderly and economically dependent population is rapidly increasing. This process is representative for Germany: Its population pyramid has been changing from an expansive towards a constrictive shape (**M2**).

The ageing population is living longer, requiring more healthcare, and often long-term care. New sheltered housing will need to be planned and built. This can be quite expensive.

These expenditures are paid for in part by the working population in Germany. The shrinking number of economically active persons aged 16-65 as compared with those economically dependent is leading to financial and social tension. Solutions to problems in the intergenerational contract are being developed. One revision is to increase the retirement age from 65 to 67.

Reasons

Germany's low birth rate is sometimes referred to as the low-fertility: a low birth rate also produces a low rate of possible parents.

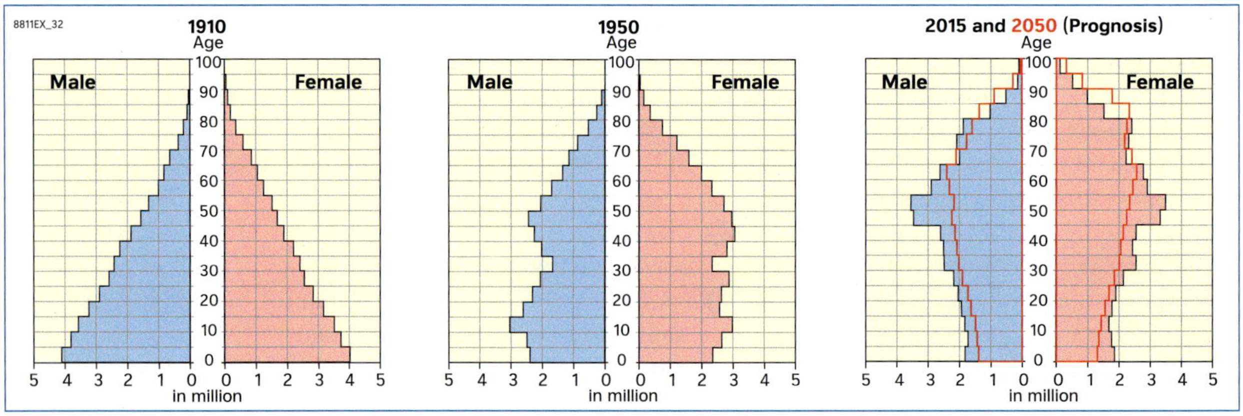

M2 Germany's population structure and scenario for 2050

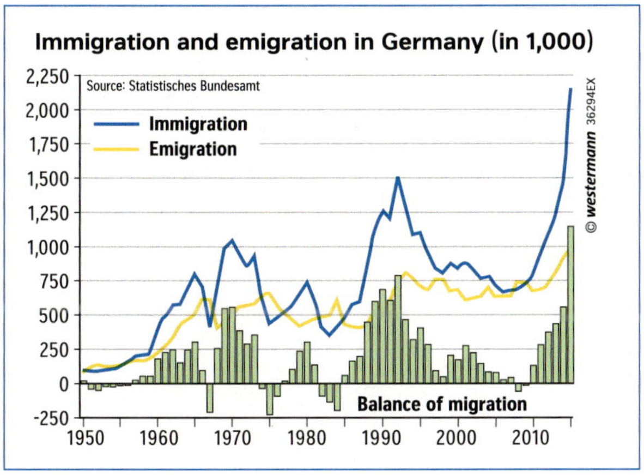

M3 Balance of migration in Germany

Important reasons for this development:

- Families in Germany are, on average, richer and more educated than in the past. They do not need children to earn money for the family.
- Since the launch of the contraceptive pill in the 1960s, family planning is easy.
- German society has become more individualised and flexible with a focus on self-fulfilment and well-being.
- Economic empowerment of women through work enables greater freedom regarding if and when to bear children.

Possible solutions

To encourage the German people to have more children, incentives were introduced, such as a cash bonus for parents ('Elterngeld'). This should help to ease the financial burden of parenthood and to make parental leave possible.
Furthermore, an increasing number of crèches and all-day schools will allow both parents to work. The expenses for daycare by public authorities have more than doubled since 1998.
Finally, immigration is an approach for Germany to balance the age structure of its population. Also, a well-directed migration policy seems to be important in order to increase the economically active population of Germany.

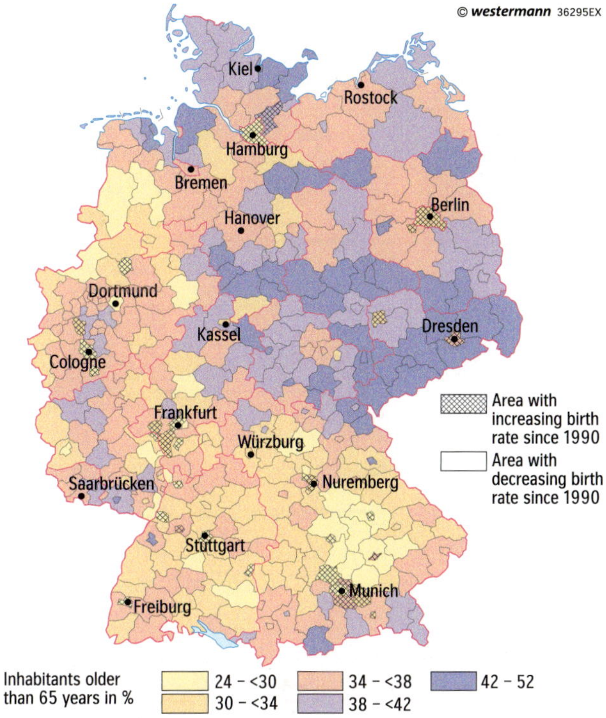

M4 Demographic trends in Germany

100800-080, -081
schueler.diercke.de

DE-042
www.diercke.com

TASKS

1. Compare the population structure of Germany in 1910 with today's (**M2**).
2. Describe the population density and age structure in Germany. Use your atlas.
3. Discuss the future challenges of demographic change in Germany. Use the **M1**, the text, and **M4**.

pp. 152 - 153 Age-Gender-Diagrams

pp. 154 -154 Cartoons

http://destatis.de/bevoelkerungspyramide/

KEY TERMS

- demographic change
- immigration
- migration policy
- rate of natural increase
- retirement age

HELPFUL WORDS AND PHRASES
for TASK 2:

- There are densely / less densely populated areas in ...

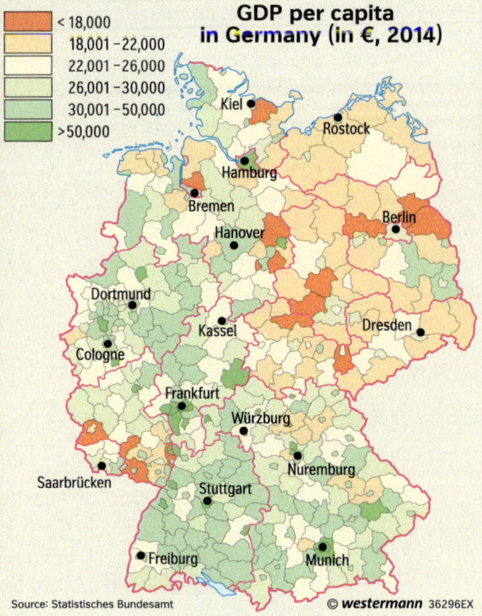

M1 GDP per capita

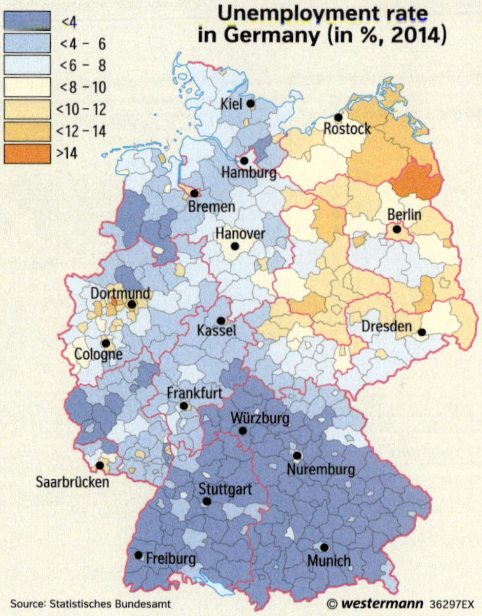

M2 Unemployment rate

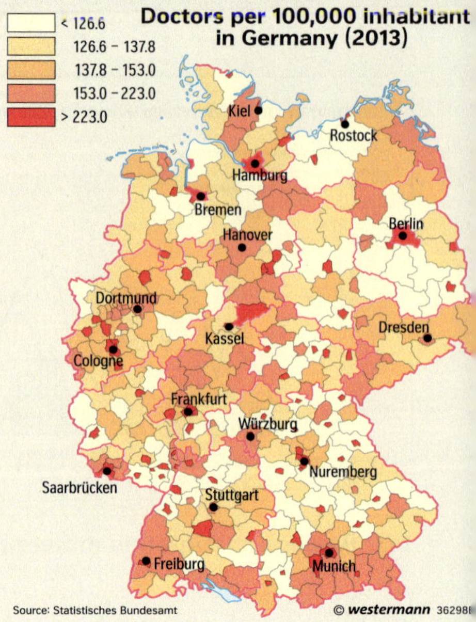

M3 Doctors per 100,000 inhabitants

A Land of Contrasts – Regional Disparities in Germany

A Ruhr deal

Cracked streets, silent factories, idle[1] workers. Eastern Germany circa 1992? Yes, but also parts of western Germany today. The Graf Bismarck coal mine in Gelsenkirchen, once one of Germany's biggest, closed in 1966. Like many of its neighbours in the Ruhr region, the city has not recovered from the near-extinction[2] of coal and steel production. Its unemployment rate is 14.6%. Nearly one-third of its inhabitants receive benefits[3]. Decades of coal-mining subsidies[4] probably delayed the Ruhr's transformation.

Gelsenkirchen is gamely[5] trying new things, and has cut its unemployment rate from still higher levels. The Graf Bismarck land was expensively reclaimed for housing and business investment, but rents are forbiddingly[6] high. Gelsenkirchen's advancing solar-energy ambitions took a knock[7] in February when Scheuten Solar, a Dutch owned company, filed for insolvency[8].

Germany 'has to live with the fact that some regions prosper and others are peripheral[9]', says Ms Günther of the Halle Institute for Economic Research. A Ruhr comeback is possible, but still looks decades away.

Source: The Economist (31 Mar 2012, abridged)

M4 Magazine report

Regional disparities exist if two or more regions differ greatly in terms of their natural conditions, population density, infrastructure, economic power, etc.

In Germany, there are regional disparities between the north and south, and in particular between the former West and East German states (M1-M3, M6). In eastern Germany, the transition to a market-oriented economy after the collapse of the German Democratic Republic (GDR) in 1989 resulted in the closure of many industries. This has led to a significant unemployment rate, which is still remarkably higher than in western Germany. Statistics show that the eastern regions have often been lagging behind, especially with respect to industrial, urban and infrastructural development.

In addition, there are also considerable disparities within the old and new federal states which are often linked to the restructuring and conversion of industrial areas. In both eastern and western Germany, rural areas are weaker than urban ones (M1-M3).

M5 Toys logistics centre in an industrial park in Sonneberg (Thuringia)

100800-070, -071 schueler.diercke.de

pp. 32 - 33 Texts

East German town shows 'flourishing landscapes' within reach at last

The eastern German town of Sonneberg has undergone something of an economic miracle. Since 1990 about 100 firms have flocked[1] to the town of 24,000 people. Economic output has grown more than fourfold and unemployment, which hit almost 13 percent by 1997, has dropped to 3.5 percent, the lowest in the east and rivaling[2] even the most prosperous areas of western Germany.

'Back in 1990 the streets of Sonneberg were in a state of disrepair[3], the houses in the town center were crumbling and some firms weren't economically viable[4]. That has all been remedied[5] now and there are hardly any differences when compared with a town in the former West Germany', Sonneberg's mayor Sibylle Abel says.

The town is not yet typical, as it has grown far faster than the rest of eastern Germany. But overall, the east is gradually catching up.

Sonneberg, which before World War One produced a fifth of the world's toys, quickly saw reunification as an opportunity. It secured subsidies[6] to improve its infrastructure and built five industrial parks, four of which are full.

Alongside a small number of surviving toy firms, the town is full of factories processing plastic, metal and ceramics and making machines and car parts.

On the outskirts[7] lies a complex of factory and warehouse buildings run by Mann+Hummel, the area's biggest industrial employer. It supplies components to big names in the car industry. 'Since reunification we've made a huge leap forward technologically and in terms of our buildings and the qualifications of our employees, so you can certainly say we've got a good basis for 'flourishing landscapes' here now', said Thomas Lange-Stalinski, head of Mann+Hummel's Sonneberg factory.

Source: www.reuters.com (Michelle Martin, 3 Nov 2014)

M6 Economic boom in Sonneberg (Thuringia)

HELPFUL WORDS

for M4:
1 – beschäftigungslos
2 – Untergang
3 – Unterstützungsleistungen
4 – Zuschüsse, Subventionen
5 – tapfer
6 – abschreckend
7 – einen Rückschlag erleiden
8 – Zahlungsunfähigkeit
9 – rückständig, am Rand

for M6:
1 – strömen
2 – in Konkurrenz mit
3 – Baufälligkeit
4 – rentabel
5 – in Ordnung bringen
6 – Zuschüsse, Subventionen
7 – Stadtrand

TASKS

1. Describe regional disparities in M1 - M3 in two to three sentences.
2. Compare the present economic development of Gelsenkirchen to that of Sonneberg.
3. Analyse the factors which have contributed to the economic boom in the town of Sonneberg.

KEY TERMS

- regional disparity
- rural area
- transformation
- unemployment
- urban area

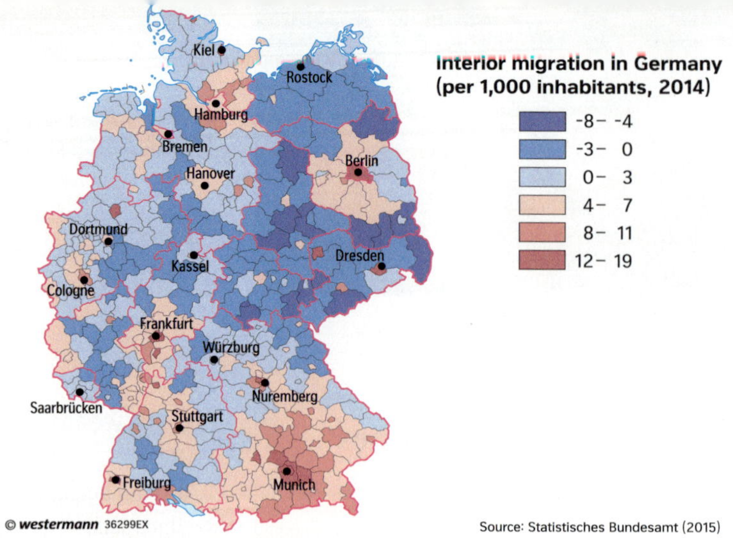

M1 Migration balance per 1,000 inhabitants

M3 Leipzig is attractive and a top-spot for in-migration in Germany

Regional Migration in Germany

People tend to move away from less developed and less attractive regions to more prosperous ones. Such movement of people within a country is called regional or interior migration. Every year about 3.5 to 4 million people move from one place (out-migration) to another (in-migration) in Germany. Apart from personal reasons, economic inequalities are a decisive factor. People move, for example, because they are looking for better job opportunities, a higher living standard, and better quality of life.

Major migration movements

Since the reunification there was a significant movement of people from East to West. Over the past few years, however, this migration flow has decreased. Today a notable number of people also move from the West to the East (**M1**, **M2**). Another migration movement is from the North of Germany to the economically better-off South.

People also tend to move away from rural areas to urban areas in order to have access to better-paid jobs and to better facilities such as hospitals, schools and the general excitement of city life. As a result, some rural areas are suffering a decline in population.

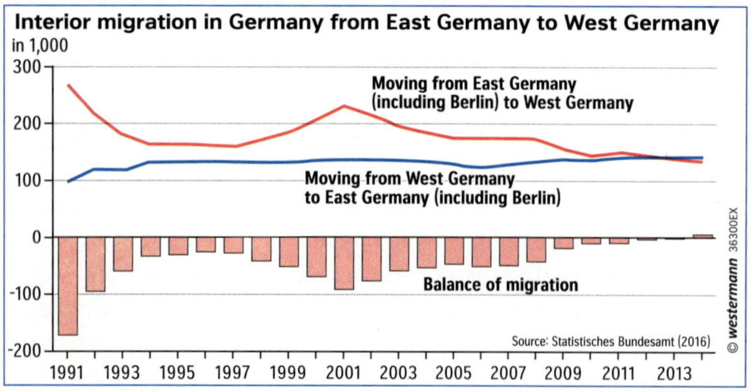

M2 Migration flows between West and East Germany

This development is especially evident in East Germany but can also be observed in some parts in the West.

However, there is also an opposing trend. As large numbers of people move into urban areas, life in the city becomes expensive and less attractive due to pollution, heavy traffic and lack of space. Those who cannot afford life in the city anymore or need more immediate nature move to the urban fringes or to the city's rural environs. This process is called suburbanisation.

The urban fringes strongly benefit from migration as they attract people not only from rural areas, but also from the inner cities. Improved transport makes commuting into the city possible (› pp. 144-1 47).

Shrinking cities

Some urban areas are experiencing a strong decline in population. At the moment, shrinking cities are concentrated in East Germany and the Ruhr area. Over time, as Germany's overall population declines, shrinking cities will become widespread across the country.

When the percentage of elderly people rises, there is little investment in growth industries because these industries need a young and highly skilled workforce. In addition, little investment occurs in the infrastructure. Such regions are trapped in a downward spiral (**M6**).

M4 Dessau – tearing down empty residential buildings

Year	Inhabi-tants
1990	114,130
1995	107,698
2000	100,029
2005	92,339
2010	86,906
2015	82,919
2030*	68,380

M6 Population development of Dessau-Rosslau (*estimated)

Age group	1995 (in %)	2015 (in %)
0 – 14	14.6	10.5
15 – 64	69.6	59.6
65 and above	15.4	25.1

M7 Population of Dessau-Rosslau by age groups 1995 and 2015

When Less Is More

Since Germany was reunified in 1990, Dessau, in the state of Saxony-Anhalt, has shrunk more than almost any other municipality[1] in the land. Whole neighborhoods appear abandoned[2], the facades of 19th century buildings are crumbling, and the windows in many of the prefabricated apartment blocks are broken. Schools, businesses and residential buildings[3] lie empty. 'What can we do when so many people leave?' city planner Heike Brückner asks. 'After all, we can't just close the town down.' [...]
No other state has been as affected by population decline as Saxony-Anhalt has. People have simply moved away. [...] Most importantly, it is mainly young people and the well-qualified who are seeking their fortunes elsewhere. The situation presents a dramatic challenge for cities. Since 2003, Brückner and her colleagues have been coming up with new strategies for coping with the shrinking cityscapes of Saxony-Anhalt. One idea is that of 'city islands' in Dessau-Rosslau. According to the concept, Dessau-Rosslau would abandon the model of a more compact central city, leaving only islands of houses. 'Buildings will be cut out and in the empty spaces we will insert[4] countryside', Brückner explains. The first results of that proposal are now visible.
Where old butchery buildings once stood, grass is now sprouting. Young people ride their BMX bicycles around a home-made course. In a dozen other bare patches, residents have planted small herb gardens or flower beds. The town's built-up area has already been reduced by over 60,000 square meters, and a further 30,000 square meters will also be demolished[5]. Two schools have already fallen victim to the wrecker's ball[6] as have 3,500 apartments.

Source: www.spiegel.de (Rainer Müller, 09 Apr 2010, abridged)

M5 Magazine report

100800-071 schueler.diercke.de

pp. 16 - 17 *Tables*

pp. 32 - 33 *Texts*

HELPFUL WORDS
for M5:
1 – Gemeinde
2 – verlassen, aufgeben
3 – Wohngebäude
4 – einfügen
5 – abreißen
6 – Abrissbirne

TASKS

1. a) Describe the major migration movements in Germany (**M1**, **M2**).
   b) Explain these major movements (text).
2. Analyse the population development in Dessau-Rosslau.
3. Make a flow chart visualizing the downward spiral cities like Dessau-Rosslau are trapped in (**M5**, text).
4. For experts: Evaluate the measures taken to improve the quality of life for people living in shrinking cities.

KEY TERMS

- in-migration
- infrastructure
- out-migration
- suburbanisation
- workforce

'Ich bin während des Studiums nach London gezogen. Dank meines Abschlusses habe ich hier eine Stelle bekommen. Ich weiß nicht, ob ich nach Deutschland zurückgehe.'

'I came here and found work in construction. With the money I make, I can support my family in Romania and build my own business when I go home next year.'

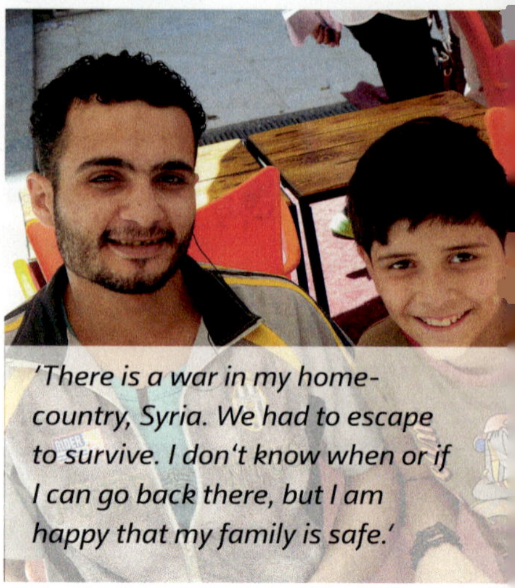

'There is a war in my home-country, Syria. We had to escape to survive. I don't know when or if I can go back there, but I am happy that my family is safe.'

M1 Migrant stories

Migration to and from Germany

It is the 1960s. In West Germany, the post-war industry is booming, and suddenly there are more jobs than workers. Konrad Adenauer, the German chancellor at the time, decides to invite guest workers from Turkey. So, after Italy, Spain, and Greece, Turkey becomes the fourth of nine countries with which Germany signs a recruitment agreement in the 1950s and 60s. The invitations extend to young workers, who are welcomed as guest workers for a limited period of time. Some of them return to their home country, but many of them stay, found families, and become an important part of German society.

M2 Guest workers in the 1960s

In 2015 and 2016, about 1.6 million people immigrate into Germany seeking asylum. Prognoses about how many will stay permanently are difficult because the Federal Office for Migration and Refugees needs some time to deal with the high number of applications.
The government points out that people who need protection should be allowed to stay. Beyond 'Welcome Centers' and temporary accomodation, those staying will need long-term integration into society and the job-market. Learning German in language classes and acquiring relevant job skills are important steps to reach these goals.

M3 Refugees in 2016

Immigration to Germany is a process that has been going on for decades and there have always been very different reasons why people immigrate. Therefore, different groups of people can be distinguished by their reasons of migration: students, migrant workers, refugees, and asylum seekers, for instance.
However, there are also people leaving Germany to return to their home countries, to find work somewhere else, or to study in a different country. Both immigration to Germany and emigration from Germany make up the country's balance of migration, which can result in either a net migration gain or a net migration loss (**M4**).

Migrant
A person who moves from one place to another, especially in order to find work.

Refugee
A person who has been forced to leave their country or home, because of a war or for political, religious or social reasons.

Asylum seeker
A person who has been forced to leave their own country because of danger, and who arrives in another country asking for permission to stay there.

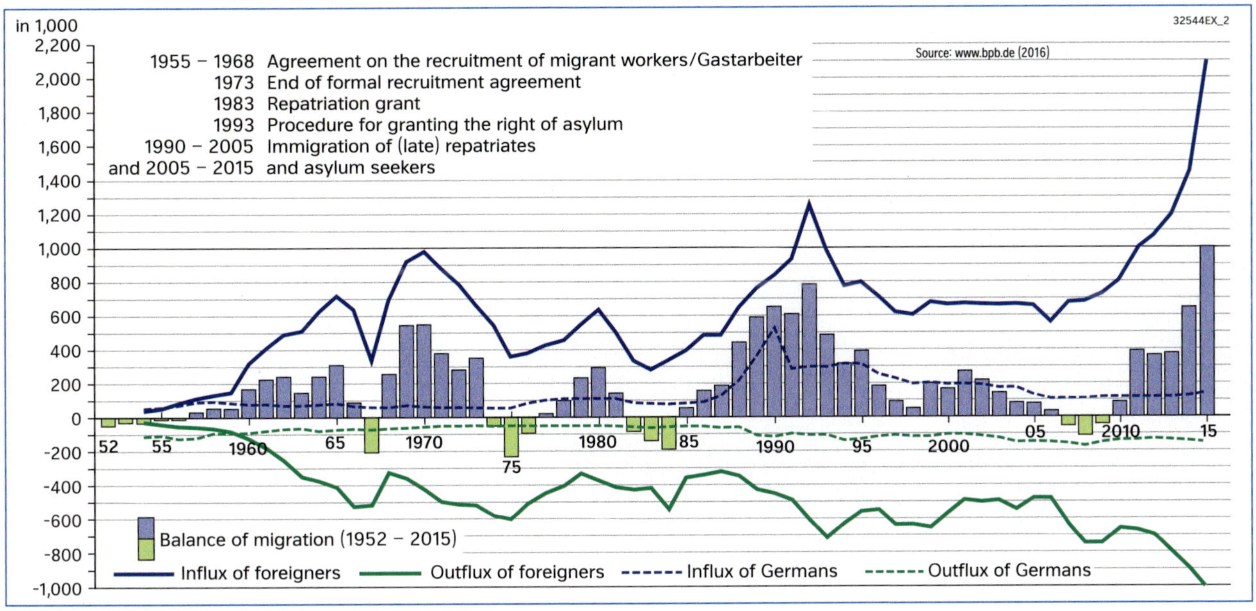

M4 Migration to and from Germany

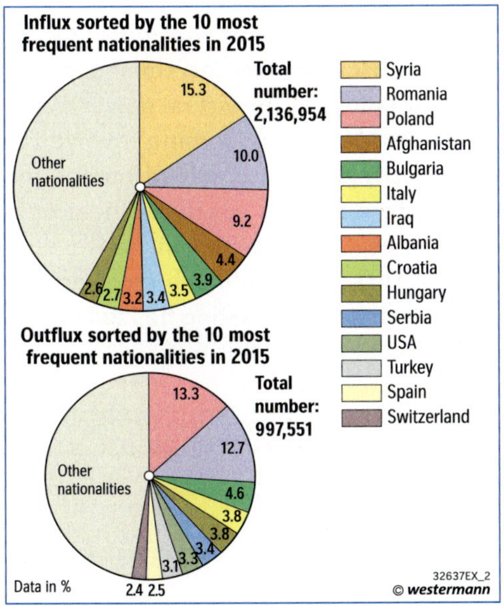

M5 Migration according to nationalities

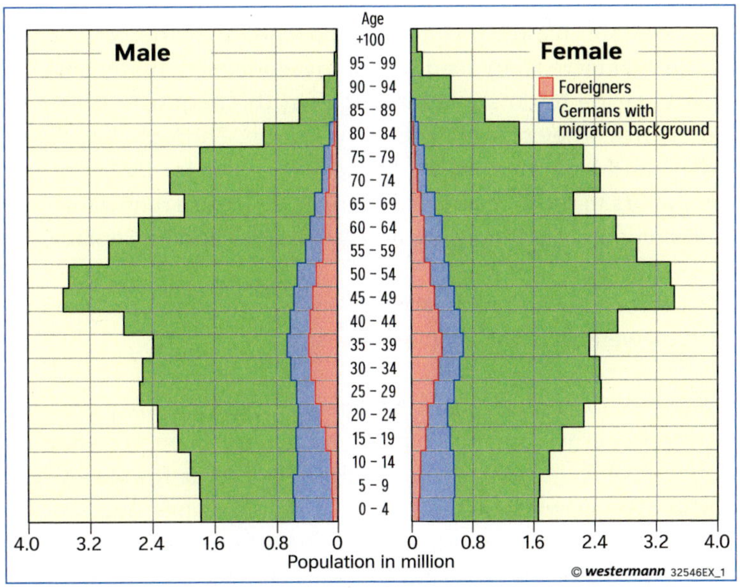

M6 Age-and-gender-diagram of Germany

TASKS

1. Describe the development of migration to and from Germany (**M4**).
2. Compare the migration of people to Germany between 1955 and 1973 as well as between 2010 and today (**M2-M6**).
3. Analyse the age structure of Germany in 2014 (**M6**).

KEY TERMS

- emigration
- immigration
- migration
- net migration gain
- net migration loss

M1 Renewable energies in Germany

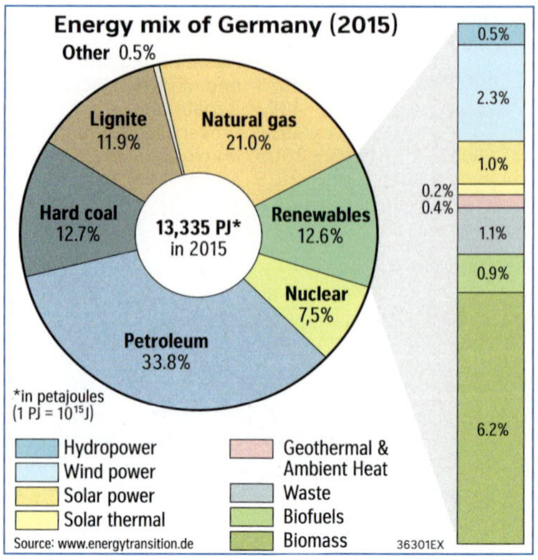

M2 Primary energy production in Germany

A Bright Future? – Germany's Energy Transition

100800-068 schueler.diercke.de

Amid climate change, pollution, and dwindling resources caused by the ever increasing need for energy worldwide, many governments decide to do something against the negative effects of energy use. They plan an energy transition. Through laws, subsidies, and taxation they try to make it easier for industry and consumers to produce and use energies which are renewable and environmentally friendly. By and large, they should replace the fossil energies.

The German government called its energy transition programme 'Energiewende'. It has six main goals:

1. *Increase the use of renewable energies.* By 2050 sixty per cent of the energy consumed in Germany should be provided by renewables. As most of the electricity will come from wind and photovoltaic which are not always available when needed, the problem of energy storage has to be addressed. One way of storing energy is by using the excess electricity to produce hydrogen or methane gas (power-to-gas, P2G) for storage. This energy conversion still loses much energy today but will become useful in the future (**M5**; › pp. 76 / 77).
2. *Increase energy efficiency.* The programme aims to obtain an even higher GDP with consuming less energy. Machines, cars, and lighting will become more efficient, houses will become better insulated against heat loss. Between 2005 and 2050 primary energy use should be nearly halved.
3. *Save energy.* The use of smart grids (› pp. 76: **M1**) will save much energy. The latest technology will be needed for a grid expansion to connect the production sites to the consumers without losing too much energy along the way.
4. *Turn electricity into the main energy source.* Households and small businesses should become virtual power plants (› p. 76: **M1**), i.e. becoming involved in electricity production to feed into the grid and store surplus energy as heat or in batteries. Combine heat and power supply (CHP), which is much more energy efficient. District heating has to become more widespread this way.

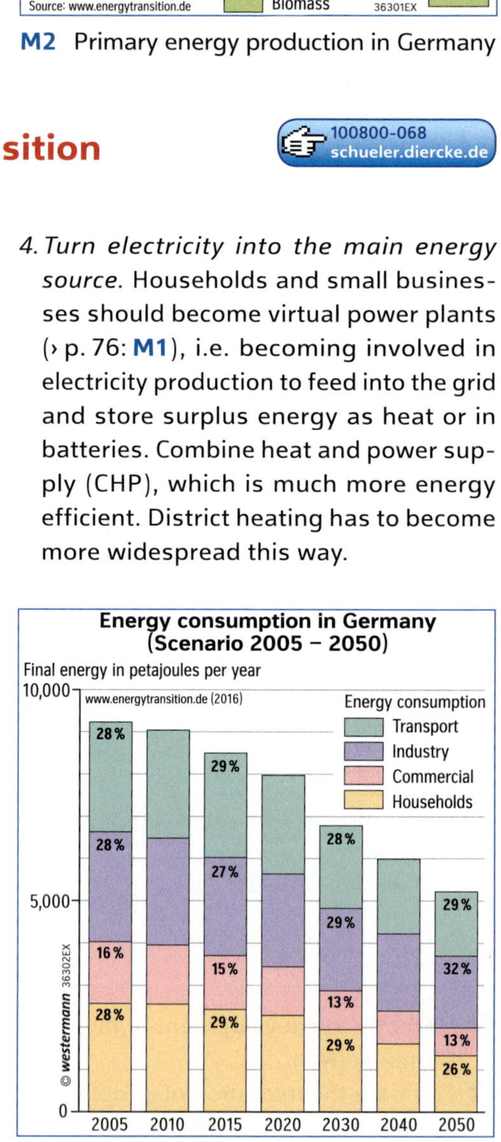

M3 Primary energy consumption in Germany by sector

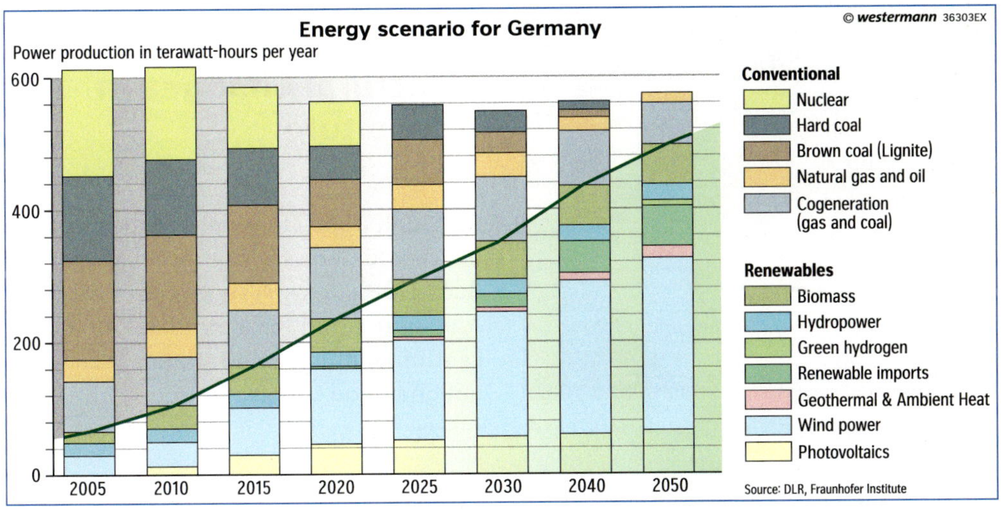

M4 'Energiewende' scenario (2005-2050) for electricity generation in Germany

pp. 26-27 *Pie Charts*

pp. 20-21 *Bar Charts*

HELPFUL WORDS AND PHRASES

for TASK 3:

- At the moment, ...
- Recently, ...
- It was suggested ...
- The main argument was ...

5. *Make use of environmental heat.* Ambient heat sources are plentiful and can be tapped with heat pumps (› p. 74).
6. *Set up an electric mobility system*, including cars, lorries and public transport on rails and roads. Charging stations have to be built for the road vehicles.

Alongside the energy transition programme, the government has decided to phase out nuclear power by 2022. The German government is confident that these goals can be reached and it has the majority of the German people behind it.

TASKS

1. Describe the 'Energiewende' programme in German.
2. Explain the importance of an energy transition.
3. Comment on the recent discussion about the 'Energiewende'.

KEY TERMS

- electric mobility
- Energiewende
- energy efficiency
- energy transition
- grid expansion
- power-to-gas

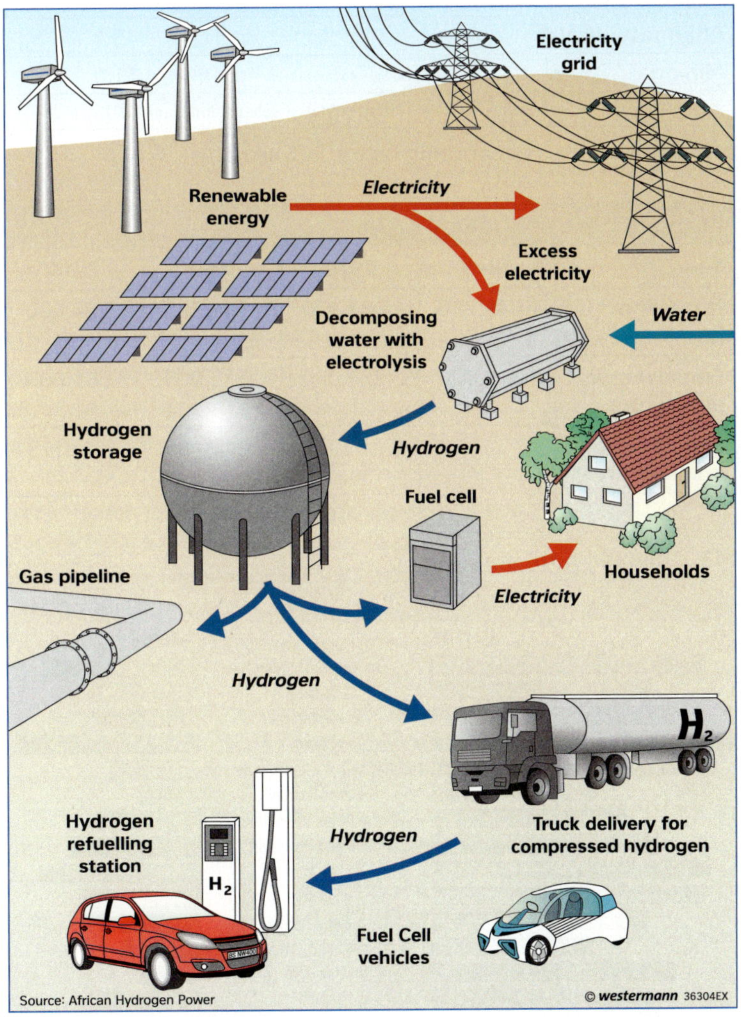

M5 Hydrogen power-to-gas storage system

M1 Agricultural land in Lower Saxony

Production value of agriculture, forestry and fishery, and net turnover of selected branches (2014, in billion €)	
Agriculture, forestry, fishery	52.7
Pulp and paper industry*	37.7
Pharmaceuticals*	45.1
Mining	11.0
Source: Statistisches Bundesamt (2016), [*2013]	

M2 Agriculture as economic factor in Germany

Changes in the agricultural sector				
Year	**1949**	**1970**	**1991**	**2014**
Number of farms	1,646,750	1,083,120	541,380	286,800
Employees in agriculture	4,819,000	2,282,000	1,535,000	651,000
People fed by one farmer	10	27	85	129*
Source: Statistisches Bundesamt (2016), [*2012]				

M3 Selected farm statistics

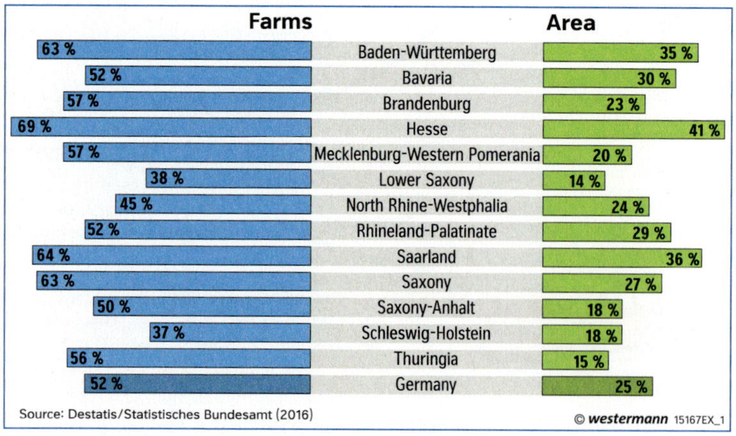

M4 Part-time farming in Germany (2013)

Structural Changes in Agriculture

Although agriculture only contributes 0.6 per cent to Germany's gross domestic product (GDP), it is still very important. Agriculture and forestry make up 83 per cent of land use, and farmers secure the national food supply.

Changes in number and size

The structure of German agriculture has changed dramatically with the number of farms shrinking as the size of those remaining grows (**M3**, **M5**). Likewise, the percentage of employees in the German agricultural sector decreased from 1950 to 2015.

The main reason for this development is a huge increase in productivity due to the development of what is called industrialised agriculture. Nowadays, agriculture is mainly characterised by the use of high-tech machinery, fertilizers and pesticides. Due to national and European agricultural policy (CAP, › pp. 114 / 115), German farmers have been under constant pressure to stay competitive. Many full-time farmers had to give up or were forced to look for a second job in other industries and became part-time farmers (**M4**).
The German reunification did not change this situation very much, except that the number of farms in eastern Germany rose as former big agricultural cooperatives had to be subdivided into smaller lots (**M5**).

Trends in German agriculture

Farming methods will continue to change considerably. Digitization, automation and robotics already play a huge role in so-called Agriculture 4.0.
Production processes are going to control themselves: machines communicate with

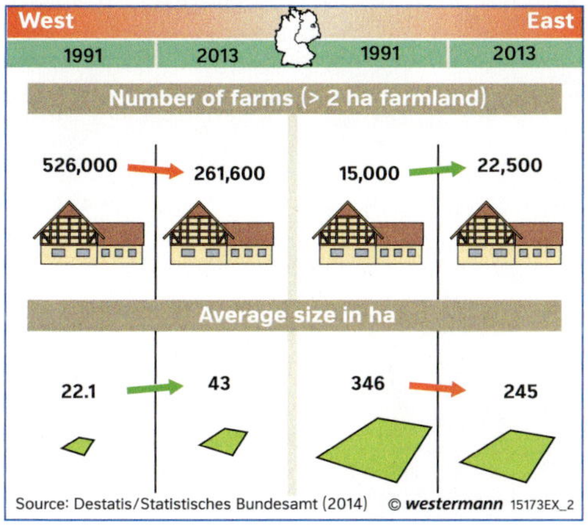

M5 Changes in farm size and numbers after the German reunification

M6 Industrialised agriculture in German (Intensive farming and automated milking system)

machines, vehicles steer independently, and computer programs make decisions. Milking systems, for example, are increasingly automated. More cows can be milked simultaneously and in a shorter amount of time while also collecting data about the quality of the milk. Feeding can also be automated, with systems making it possible to individualize fodder distribution in the stables.

Futhermore, robots, especially constructed for a certain product, can be used during harvests to speed up harvesting. The need for manual labor is also reduced.
In addition, precision farming is an important trend in modern high-tech agriculture. Here, for example, tractors with a fully automated steering system, drive across the field and take over the work of distributing the seeds, fertilizer and pesticides. During all these processes, data can be collected and used for making adjustments for the next year. In precision farming, the tractor can also harvest the crops while the farmer only drives the tractor. He sits in a fully air-conditioned driver cabin surrounded by computers and other electronic devices.
In the future, a farmer's job will be mainly focussed on programming and monitoring the automated processes.

Milk production (kg per cow)			
Year	**ca. 1900**	**ca. 1950**	**2014**
Milk yield	2155	2480	7255
Crop yield in decitonnes per hectare (dt / ha)			
Year	**1898 – 1902**	**1950 – 1954**	**2015**
Wheat	18.5	27.3	80.7
Rye	14.9	24.0	56.6
Potatoes	129.8	224.1	421.0
Sugar beets	276.8	345.5	708.0
Source: Statistisches Bundesamt (2016)			

M7 Development of yields in German agriculture

TASKS

1. List major changes in German agriculture after World War II.
2. Give reasons for the increase of the productivity (**M7**).
3. Comment on the future trends of German agriculture. Take the CAP (› pp. 114 / 115) into consideration.

KEY TERMS

- agriculture
- forestry
- industrial agriculture
- precision farming
- production value
- structural change

HELPFUL WORDS AND PHRASES

for TASK 1:

- Until the end of the 19th century, ... played an important role for ...
- With the growing importance of ...,
- ... increased steadily
- From around 1910 on ...
- Towards the end of the 20th century, ...
- Due to ..., the share of the ... sector increased.

M1 Location of the Oldenburg area

M3 Pig feeding

Case Study: Meat Production South of Oldenburg

100800-056 schueler.diercke.de

DE-046 www.diercke.com

An agricultural core region

The region south of Oldenburg, especially the districts of Vechta and Cloppenburg, is a centre of pig feeding, pig rearing and poultry farming not only in Lower Saxony but in the whole of Germany. Here you find the biggest share of the stock of pigs and poultry in Germany. Some of the big local agricultural enterprises have stocks of thousands of pigs and poultry. Many favourable conditions have supported the development of this agricultural core region (**M4**).

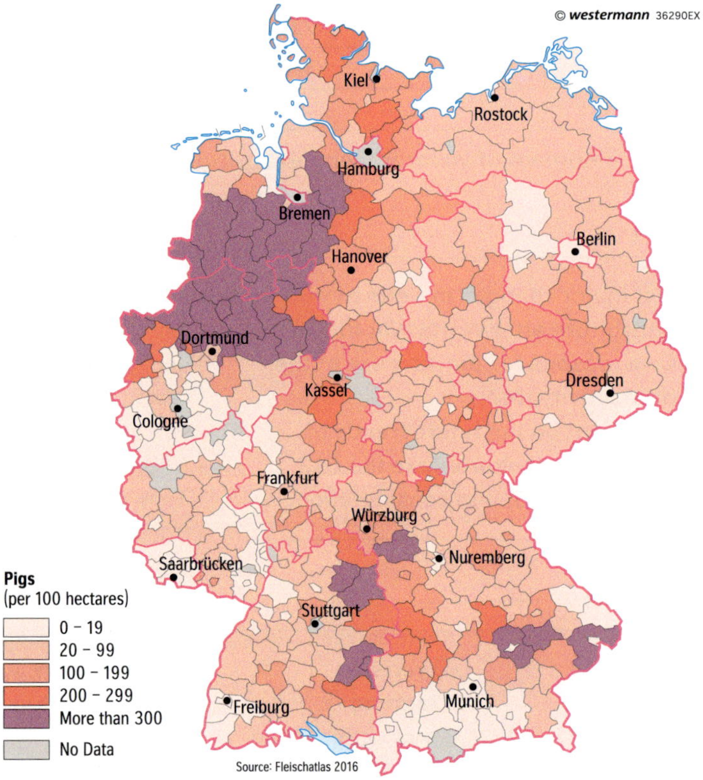

M2 Germany – pigs per 100 hectares (ha) of land

- Favourable natural conditions for the growing of fodder plants
- Well-structured transport network
- Proximity to the port of Emden
- Access to the market
- Network of slaughterhouses
- Companies specialised in the processing and marketing of meat
- Production of agricultural machinery by small and middle-sized local companies
- Maintenance service
- Agricultural research institutes
- Available labour force

M4 Location factors

Germany

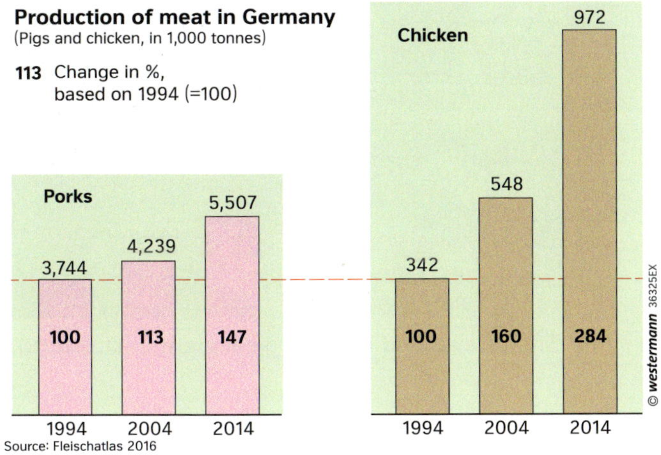

M5 Production of meat in Germany

M6 Indoor-broilers and free-range hens

Intensive livestock farming

Specialisation and modernisation have led to agricultural enterprises that have at least 2,000 pigs for feeding or 7,000 piglets for rearing to be competitive on the open market. At the same time, many small and mid-sized agricultural enterprises, mostly family farms, have given up because they lack the financial resources needed for modernisation.

As a consequence, the modern livestock-farming has changed its structure dramatically (› pp. 138 / 139). The technological progress and the economic requirements of the market have transformed the agricultural production considerably.

Small-scale livestock farming has been widely replaced by an intensive large-scale livestock farming and processing. Such intensive farming is more competitive and profitable. However, it also raises concerns about the environment and animal welfare.

Poultry farming near Oldenburg

Raising battery-caged hens is the most profitable way of feeding. However, in 2007 the law was changed. As of 2010, rearing hens in conventional battery-cages is prohibited in Germany. Instead, modified barns or so-called furnished cages are used, providing more space for the animals. Germany is planning to ban cages in poultry farming totally in the year of 2025. However, about two-thirds of all hens in the country are already raised in barns, and another quarter are free-range hens.

Intense poultry farming causes environmental problems. Due to the great number of animals concentrated on a relatively small area, the enormous amount of liquid manure creates a serious danger to the groundwater. Therefore, the authorities have decided to restrict the number of animals. As a consequence, some enterprises have left the Oldenburg area to move to places without such strict regulations.

HELPFUL WORDS AND PHRASES

for TASK 3:
- On the one hand ... / On the other hand ...
- A potential risk might be ...
- If you compare the benefits and the riks ...
- Weighing up the pros and cons ...

TASKS

1. Describe the development of the production of animal-sourced foods in Germany (**M5**).
2. Explain why meat production has become such an important economic factor in the Oldenburg region.
3. a) Explain changes in meat production in the Oldenburg region.
   b) Discuss problems caused by intensive livestock farming or poultry farming (text, internet).

KEY TERMS

- environment
- livestock farming
- modernisation
- poultry farming
- specialisation

M1 Historic image of Essen (1870)

M3 Site of a retired metallurgical plant in Duisburg

Rise and Fall of the German Coal Industry

Hard-coal mining played a major role in Germany until the middle of the 20th century. Huge industrial regions developed near local coalfields. The development began with industries producing coke and iron, followed by industries processing iron and steel. The largest of these regions became known as the Ruhr area. With a total area of about 4,400 km^2 and 5.2 million inhabitants, the Ruhr area is the biggest agglomeration in Germany and the third biggest in Europe.

The coal mining and steel crises

Starting at the end of the 1950s, the coal-mining crisis led to the closing of many coal mines (**M5**). The rise in the use of other energy resources such as crude oil and natural gas marked a turning point in Germany's coal-mining success story.

Additionally, German coal became more expensive than imported coal due to unfavourable geological conditions and high wages.

These developments were reinforced by the steel crisis: More and more products were made of plastics instead of steel and iron. Hence, less coal was needed to fire-up plants. In addition, new technologies being used in the production and processing of steel and iron reduced demand. The German government responded to these challenges by subsidising the coal industry because thousands of jobs were in danger.

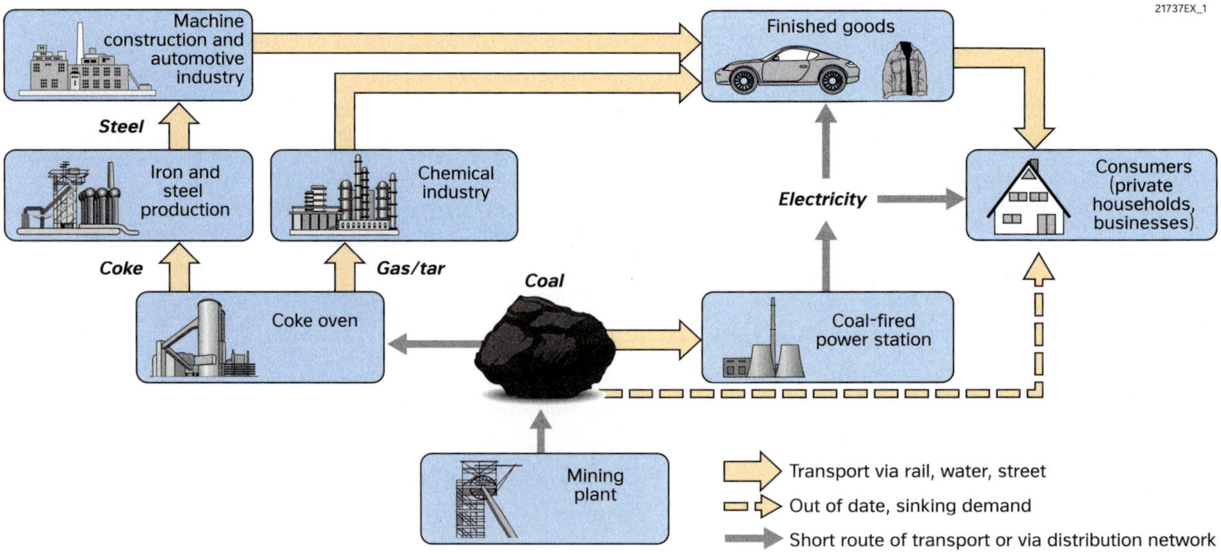

M2 Network of different industries and consumers depending on hard coal

Germany

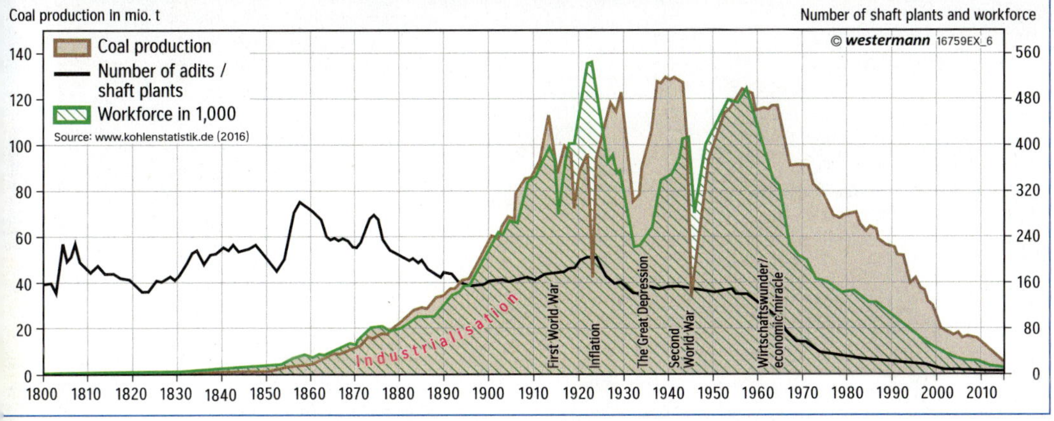

M4 Hard-coal mining in the Ruhr area between 1800 and 2010

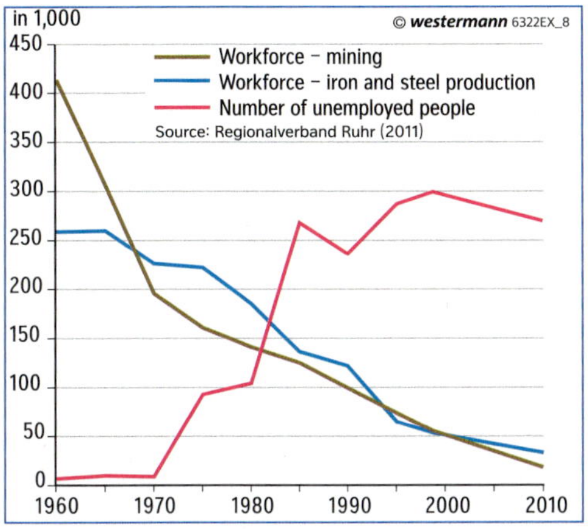

M5 Ruhr area – coal-related employment

For decades, the hard-coal mining industry in the Ruhr area has sustained generations of miners and their families. Even if the decline has been clearly visible all around Germany's former industrial heart for more than 50 years, the end of hard-coal mining signifies the extinction of a 500-year-old tradition. But German coal is too expensive. African and Asian countries are able to win coal in extremely cheap ways and ship them to Europe. The consequence is that the German government decided to let expensive subsidies expire. There were about 100 mines winning coal in the late 1950s. By 2018, the last mine in operation will be closed.

M7 2018 – the end of an area

- Mono-structured industrial network
- Large-scale operations supplying thousands of jobs
- Little basis for small-scale entrepreneurship with its flexibility and corporate readiness to assume risks
- High environmental stress due to industries / urban centers / traffic
- Disadvantages in land use and infrastructure
- Few educational and cultural institutions / locations
- Lack of growth industries and high technologies
- High amount of subsidies provided by the state

M6 Features of old industrial regions

100800-040, -067 schueler.diercke.de

DE-053 www.diercke.com

pp. 18 - 19 Line Graphs

pp. 28 - 29 Flow Charts

pp. 32 - 33 Texts

HELPFUL WORDS AND PHRASES

for TASK 3:

- Before / until now / not long ago / earlier ...
- After / afterwards / later (on) ...
- Now ...
- Due to / Owing to / As a consequence ...

TASKS

1. Describe the development of the Ruhr area between 1800 and 1960.
2. Describe the network of industries and consumers in the Ruhr area around 1960 (**M2**, atlas).
3. Explain the development of the coal industry between 1960 and today.
4. Discuss problems the Ruhr area has to face due to its features and current developments (**M5-M7**).

KEY TERMS

- coal mining crisis
- old industrial region
- steel crisis

M1 Aerial view of the mall and promenade of CentrO (Oberhausen)

Structural Change in the Ruhr Area

Besides many negative aspects, the diminishing role of the coal and steel industry also led to positive changes in the old industrial region of the Ruhr area. While, for a long time, the Ruhr area had been characterised by extreme air and noise pollution there are measures being taken to change that today: rivers are being restored, parks built, residential areas redeveloped, and former industrial sites are being converted into locations focussing on culture and education.

M2 New jobs in logistics – aerial view of Duisport (Duisburg)

New jobs in the tertiary sector have been created, especially in trade, logistics, and research. Bochum University, a keystone of the region, is one of the most important German universities in the field of research, for example. The port of Duisburg was modernised and expanded. It is the world's largest inland port today (**M2**). Finally, an old steel- and roller mill was developed and reopened as the biggest shopping centre in Europe (**M1**, **M5**).

The Ruhr Valley has been struggling against unemployment for more than 30 years now. Due to state subsidies ending in 2018, the region has to attract new business. The revenues and jobs created are needed to replace those once generated by heavy industry. However, the region needs further structural change (**M4**).

100800-040, -067
schueler.diercke.de

DE-053
www.diercke.com

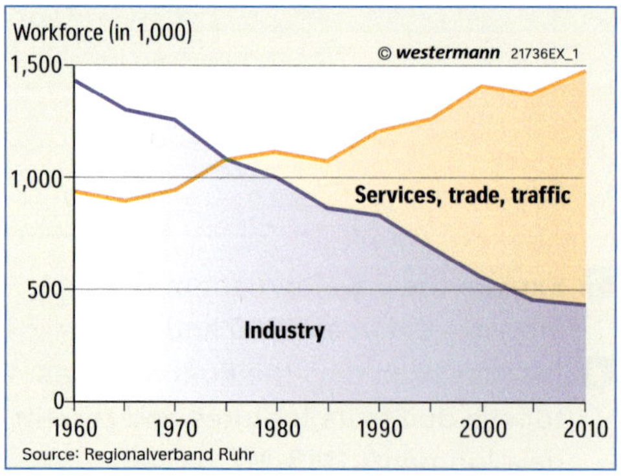

M3 Employment in the Ruhr area

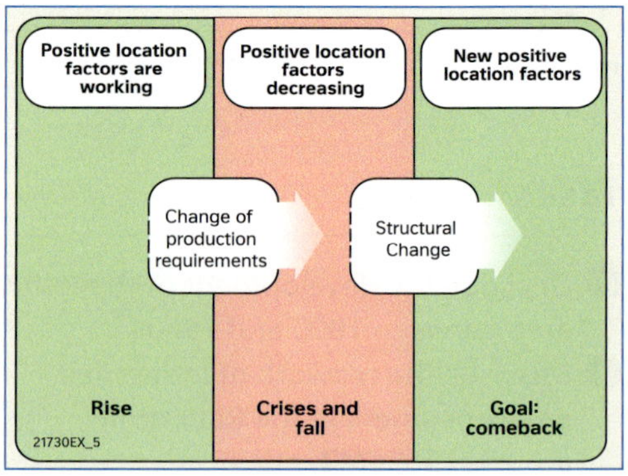

M4 Process of structural change

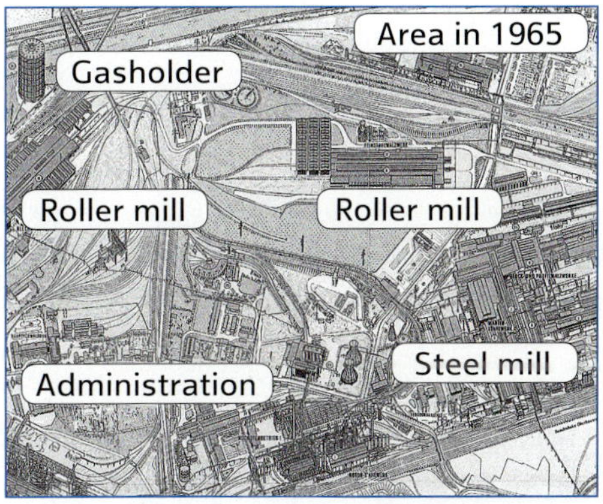

Investments in CentrO in million € (private and public: >1.1 billion €)	
Shopping centre	460
Event arena	51
Leisure park	36
Promenade with cinema	51
Business park	153
Finishings for shops, etc.	163
Infrastructure (intern)	64
Infrastructure for traffic (extern)	175
Source: CentrO Management GmbH, 2014	

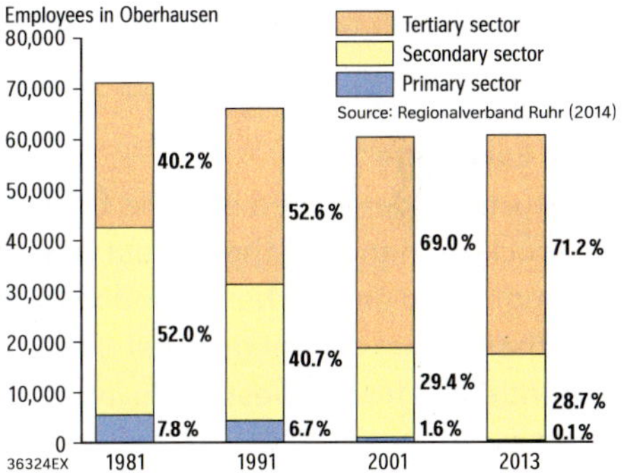

Das CentrO ist eines der größten Projekte zum Strukturwandel im Ruhrgebiet. Es besteht aus dem Einkaufszentrum, der Coca-Cola Oase, der Promenade, dem Freizeitpark und einem Business-Park. Das CentrO ist angelegt als Herzstück der sogenannten Neuen Mitte. Hier finden sich auch das Gasometer, die ARENA mit 12 000 Sitzplätzen, ein Multiplexkino, das SEA LIFE Oberhausen, das LEGOLAND Discovery Centre, die Marina und das Metronom Theater.
Über 250 Einzelhandelsgeschäfte verteilen sich im CentrO auf zwei Ebenen auf einer Nettoverkaufsfläche von 125 000 Quadratmeter. Jährlich wird das CentrO von Millionen Gästen besucht. Für Kunden mit Auto stehen bis zu 14 000 Parkplätze zur kostenlosen Verfügung. Eine eigens von der Stadt gebaute Verkehrstrasse garantiert zudem eine schnelle und komfortable Anreise mit Bus und Bahn.
Source: Centro-Daten & Fakten 11/2014, www.centro.de

M5 Information about CentrO 'Oberhausen'

TASKS

1. Describe major changes in the Ruhr area between 1980 and today (**M1-M5**, atlas).
2. Prepare a short presentation on CentrO 'Oberhausen' to an English-speaking audience (**M1**, **M5**, atlas).
3. Compare trends in employment figures and employment structure of the Ruhr Valley and Oberhausen (**M3**, **M5**).
4. For experts: Discuss whether the development of Oberhausen and CentrO is a good example of structural change in the Ruhr area (**M2**).

KEY TERMS

- coal mining crisis
- old industrial region
- steel crisis
- structural change

M1 Skyline of Frankfurt/Main

Metropolitan Region Rhine-Main

100800-044, -045
schueler.diercke.de

Frankfurt / Main is the fifth largest city in Germany with a population of about 688,000 inhabitants in 2016. It is the centre of the Rhine-Main Metropolitan region and a significant traffic junction for Germany. Frankfurt is an important economic centre and tourist destination as well because of its many historical sites, churches and museums.

Frankfurt is also of international importance. The city is the location for the headquarters of the European Central Bank (ECB). This is the central bank for Europe's single currency, the Euro. Its main task is to maintain the Euro's purchasing power – which means safeguarding the value of the Euro – and price stability in the Euro area (**M3**).

Stock Exchange

Frankfurt has developed into the financial centre of Germany. The stock exchange is the largest of the seven German stock exchanges and one of the major ones with respect to turnover. More than 50 per cent of the total trades on the Frankfurt Stock Exchange are from investors located in countries outside of Germany. Since 2010 the stock exchange has had its main office near Frankfurt in Eschborn.

M2 Frankfurt Stock Exchange

What is a central bank?

A central bank is a public institution that manages the currency of a country or group of countries and controls [...] the amount of money in circulation. The main objective of many central banks is price stability.

An individual cannot open an account or ask the central bank for a loan, because a central bank is not a commercial bank. It is not motivated by profit, (but) it acts as a bank for the commercial banks. [...] Commercial banks can turn to a central bank to borrow money, usually to cover very short-term needs.

Source: www.ecb.europa.eu (10 Jul 2015, abridged)

M3 The European Central Bank (ECB)

M4 Frankfurt Airport (FRA)

Hoechst Industrial Park

Hoechst Industrial Park is located close to the River Main. With its many firms and businesses linked to the chemical industry, it is the second largest employer behind Frankfurt International Airport. The industrial park is the former site of the Hoechst AG.

The Industrial Park has become an important site for chemical, pharmaceutical, and process industries, offering ideal conditions for research and development. Its location is ideal with easy access to international air and rail connections / transport.

M5 Industrial centre

M6 Location and traffic connections of Frankfurt Airport

pp. 12 - 13 Photographs

HELPFUL WORDS AND PHRASES

for TASK 1:

- The city of Frankfurt is important, not only because of ..., but also because ...
- Frankfurt serves as a ...
- Furthermore, it ...
- Moreover, Frankfurt is ...
- Additionally, Frankfurt offers ...

TASKS

1. Describe the importance of the city of Frankfurt for Germany.
2. Analyse the Rhine-Main region as a potential location for a company in the service industry. Use your atlas.
3. Plan a trip to Frankfurt. What would you like to visit and why?

KEY TERMS

- airline
- European Central Bank (ECB)
- hub
- industrial park
- stock exchange

M1 Leuna works – traditional centre of the chemical industry

M3 Modernised market square of Halle

Shrinkage and Growth in the Halle-Leipzig Region

The region surrounding Halle and Leipzig is a traditional economic centre in Saxony and Saxony-Anhalt. Located on the intersection of major medieval trade routes, both cities thrived. They have well-known universities, and Leipzig is important for music and publishing.
Industrialization started with the mining of lignite. This led to the construction of large chemical plants, including an ammonium plant in Leuna built in 1916. Other centres of the chemical industry were south of Halle in Schkopau, and north of Halle in Bitterfeld, and Wolfen.

After the reunification of Germany, the development of the Halle-Leipzig region has been marked by processes of both, shrinkage and growth.

Year	Population (1,000)		Unemployment rate (%)		
	Halle	Leipzig	Halle	Leipzig	Germany
1989	310	530	*	*	6.4**
1995	282	519	13.3	12.5	9.5
2000	246	493	20.9	17.7	9.6
2005	235	500	17.7	19.0	11.7
2008	231	515	16.6	14.6	7.8
2011	232	522	11.8	11.6	7.1
2013	232	531	11.5	10.3	6.9
2015	234	543	10.8	8.8	6.4
2016	237	560	10.4	8.7	6.1

Source: different sources, *no data, **only West Germany

M2 Population and unemployment

Shrinkage

The political development in Germany in 1989 led to radical changes. Both, Halle and Leipzig, lost their function as a county town. Furthermore, there was an almost complete breakdown of traditional industries in that region.
The closure of opencast lignite mines and coal processing industries increased unemployment. Since 1990 the number of inhabitants in most parts of the region has fallen. This was caused by a dramatic decline in the number of births and by migration into western parts of Germany.

In the Halle-Leipzig region, the shrinkage of urban population occurs as the suburban population on the city fringes expands. This process of suburbanisation began in the mid-1990s. By far the most prosperous suburban zone is located between the cities of Leipzig and Halle.

Suburbanisation

The process of movement of population and economic activities from within towns and cities to the rural urban fringe. It takes place primarily in more economically developed countries (MEDCs).

Germany

M4 Modernised apartment buildings in Leipzig

M5 Logistic centre near the Airport Leipzig/Halle

Growth

Change that began post-1989 in the Halle-Leipzig region is clearly evident:

- Old residential areas in Leipzig and Halle were modernized (M3).
- The infrastructure was improved:
  - motorways were reconstructed,
  - the airport Leipzig-Halle was enlarged and modernised,
  - a new trade fair site was built,
  - new logistic and service centres were opened at the same time.
- After the collapse of the old industries, a process of reindustrialisation started:
  - today, large German companies of the automotive sector have located in the region,
  - the media and the IT and communications industries provide new jobs,
  - energy- and environmental technology, centres of biotechnology, as well as research institutions are financially supported by the government,
  - the industrial park between Bitterfeld and Wolfen is a network of chemical plants, resulting from the revitalisation of traditional industrial cores,
  - Lippendorf, south of Leipzig, is the site of a modern coal-based power station.
- A number of former opencast mines have been reshaped into areas offering different opportunities for recreation (M4).

Leipzig / Halle Airport offers a wide range of outstanding advantages for airlines, as well as for air freight and logistics contractors:

- Two separate 3,600 m runways for worldwide non-stop services with no payload restrictions
- Optimum flexibility thanks to 24-hour operation for cargo flights
- Free capacities without slot restrictions
- Outstanding transport links (road, rail and air)
- Frequent passenger connections to international hubs and other European cities.

M6 Logistics centre at the Leipzig / Halle Airport

TASKS

1. Make a bar chart with the help of the figures of M2.
2. Explain the changes in the population of Halle and Leipzig.
3. Comment on the processes of shrinkage and growth in the Halle-Leipzig region since 1989.

100800-042, -043
schueler.diercke.de

pp. 16 - 17
Tables

HELPFUL WORDS AND PHRASES

for TASK 3:

- In my opinion ...
- According to ...
- As it is seen ...
- As it can be proved ...
- I want to say ...
- Let me mention ...
- The process is ...
  characterized by ...
  influenced by ...
  caused by ...
  marked by ...
  typical for ...

KEY TERMS

- infrastructure
- reindustrialisation
- revitalisation
- shopping center
- shrinkage
- suburbanisation

M1 Aerial view of Berlin with the governmental district

Berlin – The Fast Changing Capital

100800-038, -039 schueler.diercke.de
DE-037 www.diercke.com

As a result of World War II, an unusual and remarkable development took place in Berlin. Berlin was divided into West and East Berlin until 1990. Since then the capital of Germany is an attractive place of tremendous changes.

West Berlin

After the war, West Berlin was separated from the territory of the Federal Republic of Germany. Therefore, the government of West Berlin stressed cultural, economic and touristic development to keep the city attractive. Furthermore, the city needed to attract industries to settle in Berlin and create jobs.

Any development was expected to anticipate eventual reunification. Thus city planners focussed on the improvement of the transportation network. The area around the Bahnhof Zoo railway station and Kurfürstendamm gradually became the centre of West Berlin. Areas close to the Wall were usually left undeveloped.

East Berlin

Since 1949, the eastern part of Berlin was de facto the capital of the GDR. Buildings for the government, as well as for cultural, economic and educational institutions had to be built. Former places of importance, however, such as Potsdamer Platz, lost their function because they were located close to the Berlin Wall.

Changes

The city is rapidly developing (**M3**). For example, a change can be observed in the district of Prenzlauer Berg (**M5**).

This formerly run-down neighbourhood, which was populated with lower-income residents, has developed into an area where many young upper- or middle-income families, students or artists now live. Today there are various trendy cafés, pubs and clubs here.

This process, which this neighborhood has undergone since the German reunification in 1990, is called gentrification. This part of Berlin's development is reason for controversial discussions: Is Berlin made for investors or for its people?

Rents have been driven up to such levels that long-standing residents can no longer afford their apartments and have to move, usually to the city's outskirts.

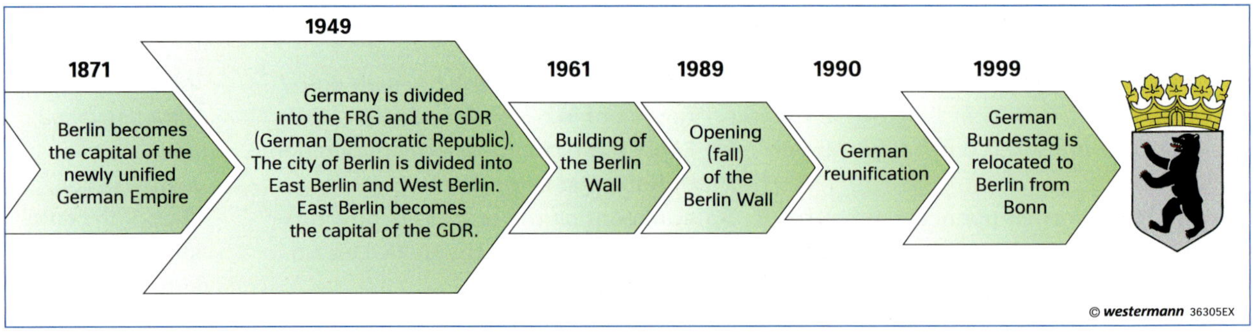

M2 Timeline of important historical events in Berlin

M3 Potsdamer Platz in Berlin around 1980 ... ... and in 2016

Berlin is cool!

- Cheap cost of living (especially for a European capital).
- Interesting job opportunities (Berlin has a world-wide renowned startup culture.)
- Berlin's music scene and nightlife ('Incredible!').
- Beautiful surrounding landscape with lakes, canals, and parks ('Europe's greenest city').
- Berlin's flair with a mix of history, architecture, governmental institutions, abandoned buildings, and street art.
- Berlin's location makes it easy to travel around Europe.
- Affordable international food is available everywhere.

M4 Why young people from foreign countries like Berlin

Swabian invasion?

'The positive side of the changes, is that literally everything looks nice now. (But) I get angry when I'm in the bakery, and there are no Schrippen (Berlin slang) only Weckern (Swabian term). I hope the Swabians realise they are now in Berlin. And not in their little towns (...). They come here because it's all so colorful and adventurous and lively, but after a while, they want to make it like it is back home. You can't have both.'

'It's not much fun going out here anymore, it's mainly wine bars, playgrounds, toy shops and yoga places, all of which isn't exactly cool, funky and exciting. (...) Prenzlauer Berg has become quite homogenized and boring (...)'

'I don't think Germans should disparage fellow Germans, wherever they are from. Especially not those footing the bills. However, I've always called it a Schrippe in the bakery!'

Source: www.spiegel.de (Matthew Tempest, 03 Jan 2013, abridged)

M5 Opinions on the gentrification process in Prenzlauer Berg

TASKS

1. Describe the changes of the Potsdamer Platz (**M3**).
2. Name four major changes Berlin has undergone since 1980.
3. Prepare a guided tour from Potsdamer Platz to the governmental district in English or in German. Use your atlas and additional material.
4. Explain the process of gentrification.

KEY TERMS

- business district
- capital
- gentrification
- government district
- government institution
- reunification

HELPFUL WORDS AND PHRASES
for TASK 2:

- Major changes can be observed in ...
- The area around ...
- The former open space has developed into ...
- The new ... which is close to ...

How to Work with an Age-and-Gender Diagram

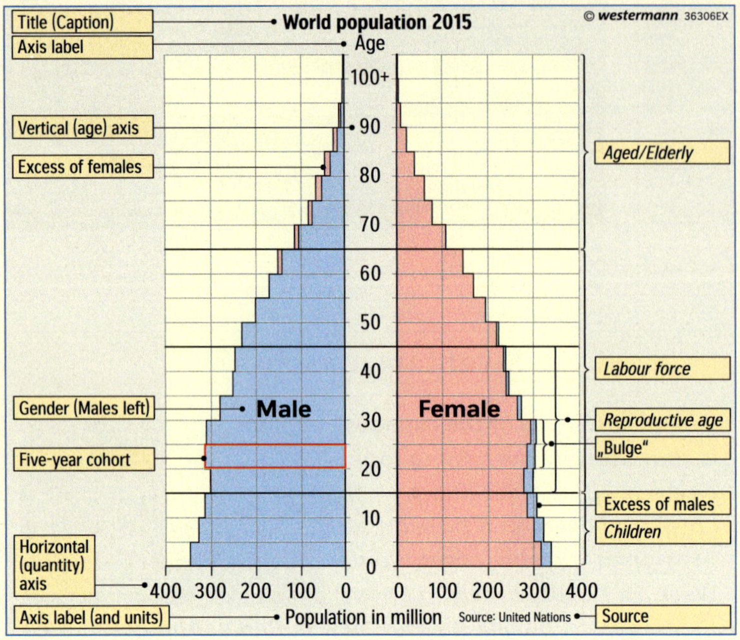

M1 Elements of an age-and-gender diagram (› p. 46: **M1**)

In an age-and-gender diagram demographic data are presented as two back-to-back histograms, one for males (on the left) and one for females (on the right). The vertical axis shows age in five-year groupings (five-year cohorts), from the youngest (0 – 4 years) up to the oldest. The horizontal axis shows either the population of the cohorts in millions or in per cent of the total population. Thus, each horizontal bar represents one age-gender cohort, for example, males between 20 and 24 years of age (**M1**).

Age-and-gender diagrams graphically display long-term trends in the birth and death rates, but also reflect shorter term baby-booms, wars, migrations, and epidemics which appear as bulges or indents in the diagram, as these incidents mostly affect certain age groups.

Steps to evaluate an age-and-gender diagram

www.worldlifeexpectancy.com

STEPS

1. Identify the overall shape of the diagram (**M2**).
2. Apply the characteristics from **M2** to your diagram.
3. Look at indents and bulges in the diagram. Use your background knowledge for explanations.
4. Give an outlook for future development of the population as an educated guess.

It is easier to compare the age structure of populations with each other when you display the cohorts as per cent of the total population. This way you can compare populations of different sizes, like, for example, China and Germany.

© westermann 36307EX

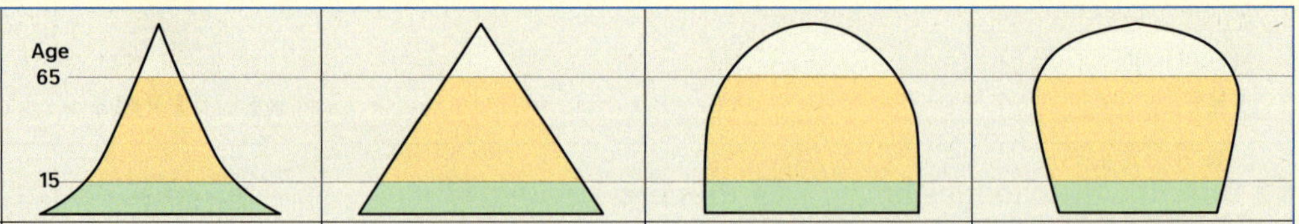

	Pagoda	Pyramid	Bell	Urn (Onion)
Birth rate	high	high	medium	low
Death rate	(very) high mortality	medium mortality	low mortality	increasing mortality
Life expectancy	very low	increasing	medium to high	high
Population growth	growing	fast growing	stationary	shrinking
Dependency ratio	low/medium	low	medium	high

M2 Basic characteristics of age-and-gender diagrams

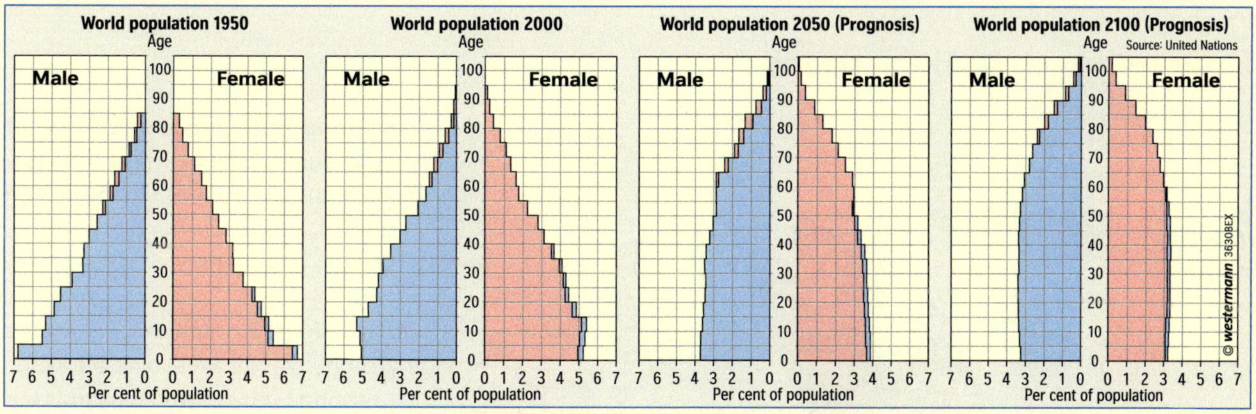

M3 Age-and-gender diagrams of the world population (1950–2100)

Model text to describe the population structure of the world in the age-and-gender diagram M1

The age-and-gender diagram shows the world population in 2015, **based on data from** the UN.

The diagram looks like a pyramid, which means that, on average, **the world birth rate is high, the death rate is medium** and **the life expectancy is increasing**. In 2015, we had a fast-growing world population with a low dependency ratio.
There is a bulge in the age group between 20 and 29 years. It is possibly due to the implementation of the one-child policy in China. So **the number of children** worldwide **may have decreased** after that, as China has a huge population.
It can be assumed that **the world population will continue to grow** in the near future, as life expectancy will increase. However, **birth rates seem to be decreasing** worldwide, so that population growth will almost certainly slow down.

HELPFUL DEFINITIONS

Birth rate: number of live births per year given in per 1000 of the population. High birth rates are above 20 per mille, low birth rates are below 10 per mille. Average world birth rate (2015): 19.73. A high birth rate leads to a wide base in the diagram. Normally, more boys are born than girls.

Death rate (mortality rate): number of deaths per year, given in per 1000 of the population. High death rates are above 12 per mille, low death rates are below 6 per mille. Average world death rate (2015): 7.74. A high mortality rate leads to big decreases from cohort to cohort.

Fertility rate: number of live births per female. For a stable population there should be 2.1 births per woman on average. World average (2015): 2.45. A high percentage of women in the reproductive age lead to a higher fertility.

Life expectancy: the average number of years that a person of a population who is born today will live. A life expectancy below 55 years is low, whereas a life expectancy above 75 is considered high. Females have a higher life expectancy than males. World average 2015: 71.7 years (69.6 males, 73.8 years females). The higher the life expectancy, the more the age-and-gender diagram looks like a bell.

Dependency ratio: the percentage of (the dependent) people below the age of 15 (children) and above the age of 65 (aged; elderly) in relation to people between 15 and 65 (labour force) who are supposed to earn the livelihood for the dependent population.

Population growth: is the result of the difference between births and deaths plus the difference between in-migration and out-migration, given in per cent of the total population per year. Worldwide it is 1.2 per cent / year (2015). There are 24 countries worldwide with negative population growth. These countries show a narrow base in the age-and-gender diagram, so that the overall shape is similar to an urn or onion.

How to Work with Cartoons

Cartoons are often used to present a strong visual message or point of view on a topic of current interest. As they are visual, rather than textual, cartoons offer lots of information at a glance.

Cartoons use a number of methods to convey their message. By exaggerating features and weaknesses they show people and things in a humorous way. Caricatures (exaggerated drawings in which some features are distorted) and symbols are often included. A symbol stands for some other thing, person, or idea, like the Stars and Stripes for the USA.

To interpret a cartoon systematically, it is useful to answer the following questions step by step, taking notes for an outline.

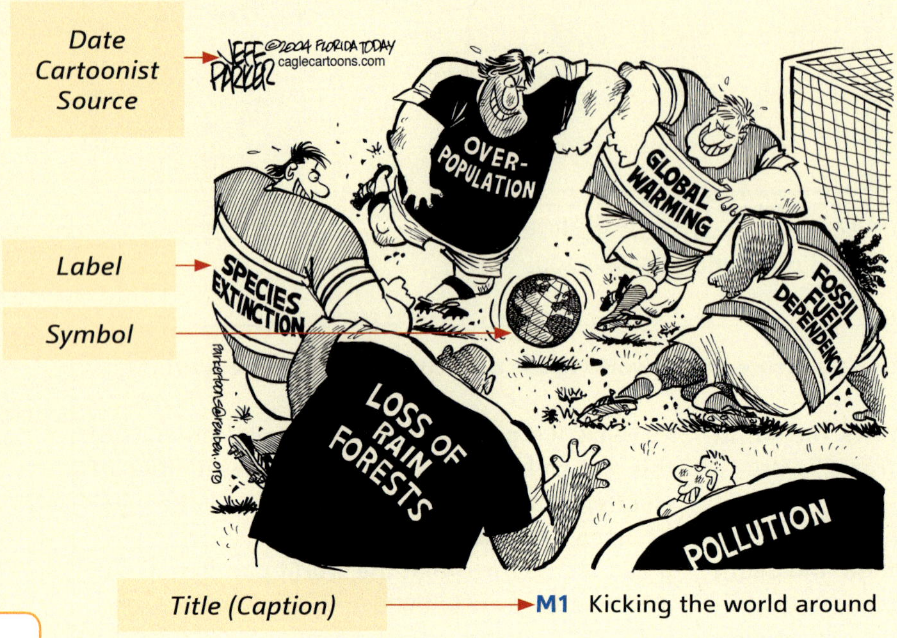

M1 Kicking the world around

M1 Elements in a cartoon (› p. 102: **M1**)

Steps to analyse a cartoon

STEPS

Step 1 – Content and source
- What are the title (caption), author, publisher, location and date?

Step 2 – Description ('What?')
- What is the event/issue that inspired the cartoon?
- What is the setting, i.e. where and when does the action take place?
- Are there any real – though caricatured – people in the cartoon?
- What is their mood, i.e. happy, sad, angry, etc.?
- Are there symbols/labels/speech bubbles in the cartoon?
- What does the cartoonist exaggerate, i.e. physical features, clothing, positioning and size of figures, facial expressions, and body language?

Step 3 – Interpretation ('Why?')
- What are the connections of the cartoon to the time and place of publication?
- What are the connections of the drawing to its title/the text of the labels/speech bubbles?
- What do the symbols stand for?
- What is the message of the cartoon?
- What is the cartoonist's intention?

Step 4 – Stating your opinion
- What is your opinion about the topic portrayed in the cartoon?
- Do you agree/disagree with the message of the cartoon?
- Are there any open questions you like to raise?

Model text to describe and analyse the cartoon M1

Source etc.

The cartoon 'Kicking the world around' **was created by** Jeff Parker **and published in** 'Florida Today' in 2004.

Description

It shows a soccer scene with two teams of three players each kicking at a ball which looks like our Earth. The players are fighting and look happy and aggressive at the same time. **This is underscored by** their stupid facial expressions and their oversized, strong bodies. The soles of their shoes have pointed spikes. The players' shirts show the words 'Overpopulation', 'Pollution', and 'Loss of Rain Forests' in one team, 'Global Warming', 'Fossil Fuel Dependency', and 'Species Extinction' in the other one.

Interpretation

The soccer players **represent** the major environmental problems of our time. **By oversizing the** players and making them violently kick against a small ball which symbolises a vulnerable Earth, **the cartoonist creates a scenario that shows** our Earth being constantly attacked and threatened ('kicked around') by these problems which are created by aggressive people who enjoy what they do.

Opinion

From my point of view, the cartoon impressively depicts the dangerous situation of our Earth. Our Earth is at the mercy of these problems. **I think it is a good idea to** personalize these problems, because they are created by people.
I share the view presented in the cartoon that we do it without much thinking and aggressively follow our own interest without regard what might happen with our planet.

HELPFUL WORDS AND PHRASES
to describe and analyse a cartoon:

Source etc.:
- The cartoon has the title '...'
- The cartoon was created by ...
- The cartoon was published in ...

Desription:
- The cartoon shows / is about ...
- In the cartoon you can see ...
- The scene depicts ...
- In the foreground/centre / background ...
- It is a caricature of ...
- The shape / size is distorted / exaggerated ...
- The gesture implies ...
- The body language shows ...
- The clothes appear to be ...
- The facial expression indicates ...
- The speech bubbles / labels say ...

- to draw attention on / to deal with / to illustrate / to represent / to symbolise ...

Interpretation:
- The message of the cartoon is / the cartoon conveys the message ...
- This is a metaphor for ...
- The cartoonist emphasises the idea ...
- The cartoonist wants to express ...
- The cartoonist's intention / message is to ...

- to suggest / to criticise / to ridicule / to expose / to make fun of / to contribute to / to stand for ...

Opinion:
- I agree in principle, but ...
- I (cannot) share the point of view ...
- As far as I am able to judge ...
- In my opinion/from my point of view ...

- (in)appropriate / (un)convincing / (un)biased / (un)realistic / exaggerated
- to support / to (dis)agree / to (dis)like / to dismiss / to reject

How to Work with some of the Most Important Instructions

Structure of a solution with the instruction 'Explain'

Note: *Explain* is a frequent exam instruction. Generally, you have to combine information from the given material with background knowledge.

Preparation
Find out what kind of explanation is asked for: Do you have to explain causes, reasons, conditions, effects, results, consequences, or purposes? Determining this is an important step to help you not to go astray with your answer.

Introduction
Repeat the topic and task in your own words
Main part
Each argument should start with a factual statement or short description, followed by the actual explanation.
Depending on the task, explain by giving a reason, by naming the causes, by pointing out the effects, etc. Link fact and explanation with a word or phrase from the wordlist. If one explanation applies to several facts, name the facts first, followed by the explanation.
Suggested paragraph structure: Fact(s) – explanation.
Link the paragraphs with **transitions** to present a coherent answer.
Conclusion
Sum up the main points and give a final comment.

Task (page 61, task 2b)
Explain the likely problems of oil transport, using your atlas.

Model text
Crude oil is transported from its production site to its consumers, mainly through pipelines and by huge oil tankers. The transport is not without dangers.
First and foremost, each leak that happens during transfer can harm the environment because oil is poisonous and flammable.
Moreover, when oil comes into contact with water it can spread easily. Thus, it can contaminate huge areas quickly and even affect the ground water.
In addition, production sites often are in politically unstable regions, so pipelines and storage tanks can become targets in warfare and terrorist attacks.
Furthermore, the atlas shows that many production sites are in remote regions of the world so that environmental damage remains unnoticed for quite some time.
Above all, transport across the oceans can be dangerous. As a look in the atlas confirms, most tankers have to pass through areas with frequent storms. When tankers are damaged or sink, then oil is released which kills the sea life around it and sticky oil residues damage coastal ecosystems.
Tourist regions suffer too, because beaches are affected by tar from oil spills.
Last but not least, oil is flammable, so there is a risk that along the transport routes burning oil will endanger human lives.
All this shows that the transport of flammable and poisonous oil endangers ecosystems and humans alike. Therefore, oil should be transported with great care.

HELPFUL WORDS AND PHRASES
to explain
... a cause or reason: as; because (of); being that; due to; for; hence; in that; in view of (the fact); owing to; since.
... a condition: as long as; even if; given that; granted (that); if; in case; in the event that; on (the) condition (that); only if; provided that; providing that; unless.
... a consequence: as a consequence; if not; if so; in that case; otherwise; that being the case; then; under those circumstances.
... an effect or a result: accordingly; as a result/an effect (of this); because of this; for this reason; hence; so; so much (so) that; so that; therefore; thus.
... a purpose: for; in order that; in order to; so (that); to the end that; with this in mind.

Structure of a solution with the instructions 'Comment on' and 'State your opinion'

Preparation
Think about the topic. What is your opinion? Jot down notes about arguments to support your ideas.

Note: Write in simple present.

Introduction
Name the problem you are going to comment on. Give a short overview on what you think.

Main part
Always give reasons for your opinion. Illustrate with examples if possible. You can comment on diverging opinions – usually in a negative way. Concentrate on one main point per paragraph. Link the paragraphs with **transitions**.
Suggested paragraph structure: Opinion – Reason (– Example).
Link the arguments.

Conclusion
Summarise and generalise the most important arguments that best support your opinion.

Task: (page 63, task 3) State your opinion on fracking.

Model text

Over the last few years, fracking has become an important means of recovering oil and gas. In my opinion it should be stopped immediately.

First of all, I fear for the environment, because there can be irreversible damages to vital water reserves. We have seen groundwater polluted by oil and gas as a result of fracking.

I have read that the supply of oil and gas leads to dropping prices. This would reduce our heating bills and transport costs. As far as I can see, that is only part of the truth, as cheap petrol and gas leads to higher consumption, which again leads to an increase of harmful pollutants and greenhouse gases. Moreover, finite resources will be depleted more quickly this way.

Finally, fracking causes earthquakes. Until now the tremors had been small, but to me this proves that fracking might trigger major quakes with catastrophic results.

There is no doubt that the environmental impact of fracking is tremendous. In my opinion, governments should advise the drilling companies to stop fracking at once.

HELPFUL WORDS AND PHRASES

... to state your opinion:
as far as I know / understand / can see, ...; from my point of view ...; I am (not) sure / certain / convinced that ...; I am of mixed opinions about / on ...; I have read that ...; I think / be-lieve / suppose ...; in my experience ...; in my opinion, ...; it seems to me that ...; my view is that ...; to me, this proves that ... .

... for linking arguments:
additionally; again; along with; also; and,; another; another aspect/point is that ...; as well; besides; equally important; finally; first(ly); first of all; further; furthermore; however; in addition; in contrast; lastly; likewise; moreover; nevertheless; next; otherwise; second(ly) etc.; similarly; too.

... for giving examples:
for example ...; for instance ...; in the case of ...; in this case; look at ...; to demonstrate; to illustrate; when / where ... .

... for the conclusion:
as (I) already indicated ...; as a result; as shown; consequently; as a consequence; finally; hence; in conclusion; in summary; on the whole; so all in all ...; summing up; therefore; thus; to conclude; to summarise; (weighing the pros and cons,) I come to the conclusion that ... .

Knowing What to Do

When you tackle a task from this textbook, you have to look out and observe instructions. They instruct you what to do when you answer the task.
In German, these instructions are called 'Operatoren'.

Some of the instructions mentioned in this book are not explained here because they are self-explaining, for example 'Write a text'. In your 'Toolkit' or on the skills pages in this book are helpful hints how to tackle most of these tasks.

Instruction	Operator	Was ist zu tun?	Beispiel	Hilfreiche Tipps
Analyse	Analysiere	Die wichtigsten Merkmale herausarbeiten. Dann im Detail – gegebenenfalls kritisch – untersuchen.	Analyse the Rhine-Main region as a potential location for a company in the service industry.	Skills S. 154-155
Comment on/ State your opinion	Kommentiere	Sachbezogen beurteilen und abschließend bewerten.	Comment on the role of renewables in the world energy supply.	Toolkit, Skills S. 157
Compare	Vergleiche/ Zeige die Verbindung zwischen	Vergleichbares nebeneinander stellen und die Gemeinsamkeiten und Unterschiede herausarbeiten.	Compare Germany's population structure of 1910 with that of today.	Toolkit
Contrast	Nenne die Unterschiede zwischen	Die Unterschiede Punkt für Punkt gegenüberstellen.	Contrast the plastic recycling industry in Dharavi with that in your home town.	Toolkit
Create/Make a diagram/ mind map/ concept map/ poster/ ...	Erstelle ein(e)	Ein Medium in ein anderes überführen. Gegebenenfalls zusätzliche Informationen sammeln.	Make a timeline that shows the changes in fishery on Lake Victoria.	Die meisten Skills-Seiten zeigen dir wie man es macht.
Describe	Beschreibe	Einen Tatbestand (Tabelle, Graphik) durch umfassende Angaben verständlich machen.	Describe the formation of crude oil and its extraction.	Es sind keine Erklärungen erforderlich.
Discuss	Diskutiere/ Erörtere	Das Für und Wider bzw. die positiven und negativen Seiten eines Sachverhaltes darstellen und zu einem eigenen Urteil kommen.	Discuss the use of solar power in LEDCs and MEDCs.	Toolkit
Explain	Erkläre	Durch Darstellen der Ursachen bzw. Zusammenhänge das Verstehen eines Sachverhaltes ermöglichen.	Explain how the terms of trade of Ghana and Nigeria have changed since 1995.	Toolkit, Skills S. 156

Hinweis:
Alle skill pages, Operatoren und weitere Hilfen zur Reihe 'Diercke Geography for Bilingual Classes' findest du gesammelt im 'Toolkit'.
(ISBN 978-3-14-114042-2)

Instruction	Operator	Was ist zu tun?	Beispiel	Hilfreiche Tipps
Give reasons	Begründe	Ursachen für einen Sachverhalt oder eine Einstellung benennen.	Give reasons for the differences in worldwide population distribution.	Toolkit
List / Make a list	Erstelle eine Liste	Eine knappe, geordnete Zusammenstellung von Fakten in Stichpunkten.	List the causes of soil erosion in the Dust Bowl.	Punkte mit Spiegelstrichen gliedern.
Locate	Verorte	Einen Standort bestimmen (Gradnetz, Atlas).	Locate the Fehmarn Belt.	Toolkit
Name	Nenne	Informationen unkommentiert wiedergeben.	Name characteristics of a global city.	Einfach aufzählen.
Prepare (for) a discussion	Bereite dich für eine Diskussion vor	Fakten und unterschiedliche Standpunkte zu einem Sachverhalt (schriftlich) zusammentragen.	Prepare for a discussion in class about your individual ecological footprints.	1. Informiere dich über die Fakten. 2. Bedenke unterschiedliche Ansichten. 3. Überlege Diskussionsstrategien.
Prepare and give a talk/ speech/ presentation	Bereite einen Vortrag/eine Präsentation vor und halte ihn (sie).	"-"	Prepare a speech, in German, about a nuclear accident.	Toolkit
Research (on)	Recherchiere	Im Internet und anderen Quellen Informationen finden und zusammenstellen.	Research which articles of your daily use are made from oil products.	Arbeite zielgerichtet und vergleiche mehrere Quellen zum Thema. Kennzeichne Zitate.
Summarize/ Sum up/ Outline	Fasse zusammen	In vollständigen Sätzen den Inhalt eines Textes knapp wiedergeben. Die Struktur nachvollziehen.	Sum up the forms of production presented on this page.	Vermeide es, Textpassagen einfach zu zitieren.

Words

A

24/7 rund um die Uhr
to **abandon sth.** etw. aufgeben, verlassen
to **abolish** abschaffen
abolition Abschaffung, Aufhebung
absorption Absorption
to **accelerate** beschleunigen
access to Zugang zu
to **account for** verantwortlich sein für
acid rain saurer Regen
to **adapt to** anpassen an
People have adapted to the natural conditions.
adequate angemessen, ausreichend
adjustment Anpassung
It might be necessary to make some adjustment to the strategy.
administration Verwaltung
advance Fortschritt
advanced fortgeschritten, hochentwickelt
advent Beginn, Einführung
to **advert to** auf etwas hinweisen
advisory board Beirat
to **advocate** befürworten, vertreten
affluence Wohlstand
to **afford** sich leisten können
affordability Erschwinglichkeit
affordable erschwinglich, bezahlbar, leistbar
afforestation Aufforstung
ageing population alternde Bevölkerung
Agenda 21 Agenda 21
agriculture Landwirtschaft
air pollution Luftverschmutzung
If there is a lot of traffic, the air pollution is very high.
airline Fluglinie
although obwohl
ambient heat Umgebungswärme
amid inmitten
amongst zwischen
apparent augenscheinlich
appliance Gerät
applicable anwendbar
application Verwendung
appropriate dazugehörig, angemessen
aquifer Grundwasserleiter, Aquifer
arable land Ackerland
to **arise** entstehen
assembly Montage
assembly line Fließband
assembly site Montageort
to **assess sth.** etw. beurteilen, einschätzen
assessment Gutachten
attempt Versuch
attendance Teilnahme
to **attract** anziehen
The Alps attract lots of tourists each year.
authority Behörde
automated automatisiert, automatisch
automation Automatisierung
Automation meant the loss of many factory jobs.
automotive industry Automobilindustrie
autonomously selbständig
availibilty Verfügbarkeit
average Durchschnitt (N.), durchschnittlich (Adj.)
to **avoid** verhindern
to **award** verleihen, zusprechen

B

backbone Rückgrat
She has the backbone to face the difficult situation.
balance of trade Handelsbilanz
to **ban** untersagen, verbieten
barrier Barriere, Schranke
basin Becken
bat Fledermaus
battery-caged in Legebatterie eingesperrt
to **be better off** wohlhabend sein, finanziell besser gestellt sein
to **be committed to do sth.** verpflichtet sein, etw. zu tun
to **be disposed of** entsorgt werden
to **be evident** offensichtlich sein
to **be observed** beobachtet werden
to **be stationary** stillstehen
to **be trapped** gefangen sein
to **bear a child** ein Kind gebären
bedrock Lagergestein
to **benefit from** Nutzen ziehen aus, profitieren von
biocapacity Biokapazität
bioenergy Bioenergie
biofuel Biokraftstoff
biogas Biogas
biomass Biomasse
birth rate Geburtenrate
bituminous coal Steinkohle
blade (Turbinen-)Schaufel
blue-collar Arbeiter-
to **boost** erhöhen, (etw.) steigern
brain drain Abwanderung hochqualifizierter Arbeitskräfte
brain gain Zuwanderung hochqualifizierter Arbeitskräfte
branch Branche
broiler Mastähnchen
burden Last, Belastung
The elderly don't want to become a burden to their children.
business district Geschäftsviertel
by and by nach und nach

C

cancellation Stornierung, Absage
capital Hauptstadt
Germany's capital city is Berlin.
to **capture** erfassen
carbon cycle Kohlenstoffkreislauf
carbon emission trading Emissionshandel mit Kohlendioxid
carbon footprint CO_2-Bilanz
cargo Ladung
catch Fang
cattleman Rinderzüchter
Central Business District (CBD) Hauptgeschäftsviertel
cereal Getreide
challenge Herausforderung
charging station Ladestation
classification Klassifikation, Einstufung
clean coal technology saubere Kohleumwandlungs-technologie
climate agreement Klimaabkommen, -vereinbarung
climate change Klimawandel
climate protection Klimaschutz
coal seam Kohleflöz
cod fishery Fang von Dorsch/Kabeljau
coin Münze
collapse Zusammenbruch
combined heat-and-power (CHP) Kraft-Wärme-Kopplung
combustion Verbrennung

Appendix

commodity Ware
commodity Verbrauchsstoff
communal forestry kommunale Forstwirtschaft
to **commute** pendeln
commuter Pendler
compatible vereinbar, passend
to **compete with** konkurrieren um
competition Wettbewerb, Konkurrenz
competitive wettbewerbsfähig
To remain competitive
compliance Übereinstimmung
The strategy is in full compliance with the law.
component Einzelteil, Bestandteil
to **compress** komprimieren
concentrated solar power (CSV) solarthermische Energie
concern Bedenken
concerning betreffend, hinsichtlich
confederation Staatenbund
confident zuversichtlich
to **confirm** bestätigen
congested überfüllt, verstopft
conserving erhaltend
considerable beachtlich, erheblich
to **constitute sth.** etw. darstellen
constrictive eng, beschränkt
consultancy Beratung
consultant Berater
consumer Konsument
consumer Verbraucher, Konsument, Abnehmer
consumption Verbrauch, Konsum
containerised containerisiert
Containerised cargo is shipped to Japan.
contamination Verschmutzung
contemplating betrachtend
continuously kontinuerlich, dauernd
contraceptive pill empfängnisverhütende Pille
to **contribute to** beitragen zu
contribution Beitrag
controversial kontrovers, umstritten
convenience food Fertigessen
convenient praktisch
conventional gas Gas aus konventionellen Ressourcen
conversion Umstellung, Anpassung
core Kern
cosmopolitan kosmopolitisch, weltoffen
crèche Kinderkrippe
In a crèche children can play with children of the same age.
to **criss-cross** kreuz und quer bewegen
crop Ernte, Getreide
crop yield Ernteertrag
cropping technique Anbautechnik
crossing Überquerung, Querung
crowded überfüllt
crucial entscheidend, wesentlich
crude oil Rohöl
currency Währung
currently derzeitig, gegenwärtig
customs duty Zoll

D

dairy product Milchprodukt
day care Kindertagesbetreuung
death rate Sterberate
debt policy Schuldenpolitik
decade Jahrzehnt
The GDP has increased considerably in the past decades.
decarbonisation Entkarbonisierung
decentralisation Dezentralisierung
decision Entscheidung
decisive ausschlaggebend, entscheidend
to **declare** verkünden, bekanntmachen
decline Niedergang
declining rückläufig, verfallend
to **deepen** vertiefen
deforestation Entwaldung, Abholzung
The area is highly affected by deforestation.
delay Verspätung
delighted erfreut, begeistert
demand Nachfrage, Bedarf
to **demand sth.** etw. verlangen, fordern
demographic change demographischer Wandel
demographic transition model Modell des demografischen Übergangs
to **demolish** abreißen
to **denote** kennzeichnen
dense dicht
A densely populated living quarter
to **depend on** abhängig sein von
dependent Abhängige(r), hier: nicht Erwerbstätige(r)
depleted aufgebraucht, abgebaut
deposit Lagerstätte
deposit Lagerstätte
to **deprive of sth.** berauben
derelict aufgegeben, baufällig
Revitalisation of derelict areas is necessary.
despite trotz
destination Ziel, Bestimmungsort
developed country Industrieland
developing country Entwicklungsland
development Entwicklung
device Gerät
diet Ernährung
to **differ** sich unterscheiden
digitization Digitalisierung
directive Richtlinie
discovery Entdeckung
to **dismantle** abbauen
disparitiy Ungleichheit
to **dispose of** entsorgen
distillate Destillat
to **distinguish** unterscheiden
district Bezirk, Kreis
district heating Fernwärme
disturbance Störung
divergent auseinandergehend, unterschiedlich, divergierend
division Aufteilung
domestically inländisch
downtown Innenstadt, Stadtzentrum
Many businesses are located downtown.
downward spiral Abwärtsspirale
drainage Abfluss, Ableitungskanal
drilling mud Bohrschlamm
driving force treibende Kraft
drought Dürre
due to aufgrund von, durch
Due to climatic conditions, settlement is not possible there.
Dust Bowl Trockengebiet
dweller Bewohner
dwelling Wohnung
dwindling schwindend

E

ecological footprint Ökologischer Fußabdruck
ecological overshoot Überschreitung der ökologischen Belastbarkeit/Grenze
economic boom Wirtschaftsaufschwung
economic sector Wirtschaftssektor

economy Wirtschaft
elderly Senioren, ältere Menschen
electric grid Stromnetz
electric mobility Elektromobilität
to **embody** verkörpern
to **emerge** entstehen
emergence Entstehung
emigrant Auswanderer
emigration Auswanderung
emission Emission, Ausstoß
to **emit** abgeben
to **employ sb.** jdn. beschäftigen
employee Mitarbeiter, Arbeitnehmer
to **empower sb.** jdn. stärken, Unabhängikeit verleihen
empowerment Ermächtigung
She worked for the empowerment of the poor.
to **emulate** nachahmen
to **encourage sb.** jdn. ermutigen, ermuntern
Energiewende
energy conversion Energieumwandlung
energy crop Energiepflanze
energy efficiency Energiewirksamkeit
energy transition Energiewende
enlargement Vergrößerung, Ausweitung
to **enrich** bereichern, (etw.) anreichern
The soil is enriched with nutrients.
to **ensure** sicherstellen, gewährleisten
enterprise Unternehmen, Betrieb
to **entrust** betrauen
environment Umwelt, Umgebung
We keep the environment clean.
environmental impact Umwelteinfluss
environmental problem Umweltproblem
environs Umland, Umgebeung
to **equal** gleichen
erosion Erosion, Bodenabtrag
to **establish** gründen, begründen, einrichten
establishment Einrichtung, Errichtung
to **estimate** schätzen
ethnic group Ethnische Gruppe
European Central Bank (ECB) Europäische Zentralbank
evidence Nachweis, Beweis
to **evolve entwickeln**
to **exceed** überschreiten
exception Ausnahme
excess (adj.) überschüssig
expansion Erweiterung, Ausdehnung
expenditure Ausgaben, Aufwand
to **exploit** ausnutzen
exploitation Ausbeutung
to **explore** erforschen
export good Exportgut
exposure Ausgesetztsein
Exposure to harmful sunlight
to **extend to** sich erstrecken auf, sich beziehen auf
extinction Aussterben
to **extract** entziehen
extraction Förderung

F

to **face** konfrontiert sein mit etw., etw. gegenüberstehen
to **facilitate** ermöglichen
facility Anlage, Einrichtung
to **fade** verblassen
fallow Brache (N.), brachliegend (Adj.)
fallow land Brachland
farming Landwirtschaft
favourable positiv, vorteilhaft
The site offers favourable location factors.
feeder traffic Zuführungsverkehr
feeding system Fütterungssystem
fertility Fruchtbarkeit
fertility rate Fruchtbarkeitsrate, Fertilitätsrate
fertilizer Düngemittel
fibre optic cable Glasfaserkabel
figurehead Galionsfigur
Modern kings are often figureheads.
finishing Fertigstellung
fiscal policy Finanzpolitik
fish capture Fischfang
fish farm Fischzuchtbetrieb
fish stocks Fischbestand
fishery Fischerei
fission Kernspaltung
flood Überschwemmung
flood protection Hochwasserschutz
to **flourish** gedeihen, aufblühen
fodder Futtermittel
foe Feind
food security Ernährungssicherheit
to force sb. to to do sth. jdn. zwingen etw. zu tun
Fordism Fordismus
foreign direct investment ausländische Direktinvestition
forestry Forstwirtschaft
fossil fuel fossiler Brennstoff
to **foster** fördern
fostering Förderung
fracking Fracking
fragementation Fragmentierung, Zersplitterung
free market economy freie Marktwirtschaft
free trade zone Freihandelszone
free-range hen freilaufendes Huhn
fresh water Trinkwasser
fringe Randzone
On the southern fringe of the city
fuel Brennstoff
fuel cell Brennstoffzelle
full-range trader Vollsortimenter
full-time farmer Vollerwerbslandwirt

G

gadget Gerät
to **gain** gewinnen, (etw.) erhalten
gainful employment Erwerbstätigkeit
garbage removal Müllabfuhr
gasholder Gasometer
gender equality Geschlechtergleichstellung
generous großzügig, wohlwollend
gentrification Gentrifizierung
geothermal energy Geothermalenergie
global city globale Stadt
global warming Globale Erwärmung
globalisation Globalisierung
goal Zielsetzung
goat Ziege
governmental district Regierungsviertel
governmental institution staatliche Institution
gradual allmählich, schrittweise
to graduate einen Schulabschluss erlangen
grant Zuschuss, Untersützung
Germany has given Ghana a grant of €10 million.
greenhouse effect Glashaus-/Treibhauseffekt
greenhouse gas Treibhausgas
grid expansion Netzausbau
gross domestic product (GDP) Bruttoinlandsprodukt (BIP)
groundwater Grundwasser

H
habitat Habitat, Lebensraum
harvest Ernte
headquarters Hauptsitz, Zentrale
healthcare Gesundheitsversorgung, Gesundheitsfürsorge
heat pump Wärmepumpe
heat sink Kühlmedium
heavy metal Schwermetall
high-quality good hochwertige Ware
high-tech industry High-Tech-Industrie
hiking path Wanderweg
homogeneous homogen, einheitlich, gleichartig
hospitality sector Gastgewerbe(-bereich)
host country Gastland
hub Drehscheibe, Knotenpunkt
The port of Singapore has developed to an important transport hub.
Human Development Index (HDI) Index der menschlichen Entwicklung
hustle and bustle hektisches Treiben, Trubel
hydro energy Wasserkraft
hydroelectric dam Wasserkraft-Talsperre

I
ice core Eiskern
illegally illegal
imbalanced diet unausgewogene Ernährung
immersed tunnel Absenktunnel
immigrant Einwanderer
immigration Einwanderung
impact Auswirkung
to **implement sth.** etw. umsetzen, durchführen
import good Importgut
to **improve** verbessern
improvement Verbesserung
impurity Verunreinigung
in particular insbesondere, speziell
in terms of in Bezug auf
incentive Anreiz, Ansporn
The gold medal was an incentive for him to improve his performance.
income Einkommen
indeed allerdings, tatsächlich
independence Unabhängigkeit
independent unabhängig
indicator Indikator
The GDP is often used as an indicator for ...
indispensable unverzichtbar
industrial agriculture industrialisierte Landwirtschaft
industrial park Industriegebiet
inequality Ungleichheit
to **infiltrate** unterwandern, eindringen
infinite unerschöpflich
influx of Zustrom von
informal settlement informelle Siedlung, Elendsviertel
infrastructure Infrastruktur
to **inhabit** bewohnen
inhabitant Einwohner
inheritance Erbe, Hinterlassenschaft
in-house production depth Produktionstiefe
inital anfänglich
inland port Binnenhafen
in-migration Zuwanderung
insecticide Insektizid, Insektenvernichtungsmittel
in-situ leaching (ISL) Untertagelaugung
insurance Versicherung
to **intend** beabsichtigen
to **intensify** intensivieren
intensive intensiv
intention Absicht, Vorhaben
interconnection Verbindung
intergenerational contract Generationenvertrag
interlocking ineinander greifend
Interlocking parts of the machine
internal migration Binnenwanderung
invention Erfindung
invisible unsichtbar
irrigation Bewässerung

J
to **jeopardize** gefährden
job vacancy freie/unbesetzte Arbeitsstelle
joint venture Zusammenschluss von Unternehmen
Shell has founded joint ventures with the Nigerian National Petroleum.

K
know-how Fachwissen

L
labour force Arbeitskraft
labour migration Arbeitsmigration
to **lack** fehlen, ermangeln
to **lag behind** zurückbleiben, hinteherhinken
land clearance Rodung
land reclamation Landgewinnung
landed gentry niederer Landadel
legislation Gesetzgebung
leisure amenities Freizeitangebot
Less Economically Developed Country (LEDC) Entwicklungsland
life expectancy Lebenserwartung
In MEDCs people have a high life expectancy.
light-flooded lichtdurchflutet
lignite Braunkohle
limitation Einschränkung, Begrenzheit
linkage Verbindung
liquefied verflüssigt
liquefied natural gas (LNG) verflüssigtes Erdgas
literacy rate Alphabetisierungsrate
livelihood Lebensgrundlage
Livelihoods can be threatened by severe thunderstorms.
livestock farming Tierhaltung, Viehwirtschaft
Local Agenda 21 Regionale Agenda 21
local council Gemeinderat
location factor Standortfaktor
to **log** fällen, abholzen
They logged the forest to enlarge the building area.
logging Abholzung
logistic company Spedition
low-cost carrier Billigflieger

M
magnetic levitation train Magnetschwebebahn
to **maintain** aufrechterhalten
maintenance Wartung, Instandhaltung
malnutrition Unterernährung,Mangelernährung
manifold vielfältig
to **manufacture** herstellen
manufacturing Herstellung, Fertigung, Produktion
manure Gülle
marine energy Meeresenergie
marine pollution Meeres-/Ozeanverschmutzung
market share Marktanteil
maternal health Müttergesundheit
mature ausgereift, reif
means Mittel
means of transport Transportmittel
Mediterranean Mittelmeer
to **meet challenges** Anforderungen bestehen
melting pot Schmelztiegel
Big cities often become a melting pot of cultures.
to **mention** erwähnen
metabolism Stoffwechsel
metallurgical plant Hüttenwerk

methane Methan
methane hydrate Methanhydrat
metropolitan area Metropolregion
micro-hydro power station Mini-Wasserkraftwerk
migrant worker Wanderarbeiter
to **migrate** wandern, migrieren
to migrate from ... to ...
migrating bird Zugvogel
migration Wanderungsbewegung; Migration
migration flow Migrationsstrom
migration policy Migrationspolitik
milking system Melksystem, -Anlage
minimum energy performance standard (MEPS) Mindestnorm für Energieeffizienz
modernisation Modernisierung
Agriculture is undergoing modernisation.
monetarily finanziell
monetary union Währungsgemeinschaft, Währungsunion
monitoring Überwachung
monolithic monumental
More Economically Developed Country (MEDC) Industrieland
mortality Sterblichkeit
mortgage lending company Hypothekargeschäft
multinational company multinationaler Konzern

N

national economy Volkswirtschaft
native einheimisch
natural gas Erdgas
natural increase natürliches Bevölkerungswachstum
net contributor Nettozahler
Newly Industrialized Country (NIC) Schwellenland
non-conventional gas Gas aus unkonventionellen Ressourcen
non-governmental organisation (NGO) Nichtregierungs-organisation
non-renewable energy nichterneuerbare Energie
nuclear disaster Atomunfall
nuclear energy Kernenergie
nuclear fuel cycle Kernbrennstoffkreislauf
nuclear power station Kernkraftwerk
nutrient Nährstoff
nutrition Ernährung

O

to **observe sth.** etw. beobachten, wahrnehmen
obsolete überflüssig
obvious offensichtlich
occupation Beschäftigung, Beruf
to **occur** sich ereignen, auftreten
ocean current Meereströmung, -strom
octagon Achteck
offshore auf offener See
oil crisis Ölkrise
oil shale Ölschiefer
oil spill Ölverschmutzung
oil well Ölbohrloch
onshore an Land
OPEC OPEC
open pit mining Tagebau
opposing entgegengesetzt
to **originate** entstehen
to **outdistance** überholen, abhängen
out-migration Abwanderung
outskirts Außenbezirk, Randgebiet
To live in the outskirts of a city
outsourced ausgelagert
outsourcing Outsourcing, Auslagerung
Some companies outsource parts of their production.

to **overcome sth.** etw. bewältigen
overcrowded überfüllt, überlaufen
overfishing Überfischung
to **overgraze sth.** etw. überweiden
to **owe so. sth.** jmd. etw. verdanken

P

parental leave Elternzeit
particulates Feinstaub
part-time farmer Nebenerwerbslandwirt
to **pass** passieren, durchlaufen
pension Rente
to **perceive** wahrnehmen
peril Gefahr
periphery Peripherie, Stadtrand
Industrial development on the periphery of the town
persistent ständig, anhaltend
pesticide Pflanzenschutzmittel
petrochemical industry Petrochemie
petrol Benzin
pharmaceutical arzneilich, pharmazeutisch
to **phase sth. out** etw. schrittweise beenden
photovoltaic (PV) Photovoltaik
piglet Ferkel
pipeline Pipeline
plastics Plastik, Kunststoff
to **plough** pflügen
pneumonia Lungenentzündung
politician Politiker(in)
pollutant Schadstoff
pollution Umweltverschmutzung
population density Bevölkerungsdichte
population distribution Bevölkerungsverteilung
population growth Bevölkerungswachstum
potent wirksam
poultry Geflügel
poultry farming Geflügelzucht, -Haltung
poverty Armut
Many elderly people live in poverty.
poverty immigration Armutseinwanderung
power station Kraftwerk
power-to gas (P2G) Power-to-Gas
prairie Steppe
precision farming Präzisionslandwirtschaft
predator Raubtier, Raubfisch
to **predict** voraussagen
predominant vorherrschend
to prefer vorziehen, begünstigen
pregnancy Schwangerschaft
to **preserve** bewahren
Efforts to preserve the peace have failed.
prevalent verbreitet, vorherrschend
prevention Vermeidung
primary energy Primärenergie
primary sector primärer Sektor
privilege Privileg, Vorrecht
to **process** verarbeiten
processing Verarbeitung
production site Produktionsstätte, -Standort
production value Produktionswert
productivity Produktivität
profitable profitabel, gewinnbringend
progress Fortschritt
to **prohibit** untersagen, verbieten
prosperity Wohlstand, Wohlergehen
prosperous wohlhabend, finanziell besser gestellt
to **provide sb. with sth.** jdn. mit etw versorgen
provision of Bereitstellung von, Versorgung mit
proximity Nähe
pumped-storage power station Pumpwasserkraftwerk

purchasing power Kaufkraft
purpose Zweck
to **pursue** verfolgen
She wishes to pursue a medical career.

R

radioactive waste Atommüll
rate of natural increase natürliche Wachstumsrate
readiness Bereitschaft
to **reaffirm** nochmals bestätigen
to **reappear** wieder erscheinen
to **rear** aufziehen
recent vor kurzem, jüngst
to **reconcile sth. with sth.** etw mit etw vereinbaren
reconciliation Versöhung, Wiederaussöhnung
to **reconsider** überdenken
recovery Förderung
recreation Erholung
to **reduce** reduzieren, verringern
to **refer to** sich beziehen auf
reform Reform
refugee Flüchtling
to **regulate** regulieren
The government regulates the amount of water used.
regulation Vorschrift, Bestimmung
reindustrialisation Reindustrialisierung
to **reinforce** verstärken
reliable verlässlich
Agriculture is a reliable source of income.
relocation Verlagerung, Standortwechsel
to **rely on sth.** auf etw. beruhen
remaining verbleibend, restlichen
remarkable bemerkenswert, auffallend
remarkably auffallend, außergewöhnlich
remote abgelegen
removal Beseitigung
renewable energy erneuerbare Energie
rental price Mietpreis
to **replace** ersetzen
reprocessing facility Wiederaufbereitungsanlage
reputation Ruf
to **require** benötigen
requirement Anforderung, Voraussetzung
research centre Forschungszentrum
research facility Forschungseinrichtung
to **resemble** ähneln
reserve Reserve
resettlement Umsiedlung
resident Bewohner, Anwohner
The local residents got angry because of the noise.
residential area Wohngebiet
residential buildings Wohngebäude
residue Abfall, Rückstand
resistance Widerstand
resource Ressource
to **restrict** beschränken, einschränken
to **restructure** umstrukturieren, neu ordnen
restructuring and conversion of industrial areas
Umstrukturierung und Umbau von Industriegebieten
retired stillgelegt
retirement Ruhestand
retirement age Rentenalter
reunification Wiedervereinigung
revenue Einnahmen
The company's annual revenues rose.
revision Neuregelung
revitalisation Revitalisierung, Wiederbelebung
roaming charge Roaming-Gebühr
roller mill Walzwerk
rooftop Hausdach
roughly ungefähr
run-down heruntergekommen
run-off Abfluss
Run-off of rainfall
run-of-the-river power station Laufwasserkraftwerk
rural ländlich
rural exodus ländliche Abwanderung
rural to urban migration Land-Stadt-Wanderung
rye Roggen

S

salary Gehalt
saleable verkäuflich
salinisation Versalzung
sanitation sanitäre Einrichtungen
to **satisfy** zufriedenstellen, befriedigen
to **schedule** planen, ansetzen
seal Dichtung
seasonal worker Saisonarbeiter
secondary energy Sekundärenergie
secondary sector sekundärer Sektor
seed Saat, Samen
segregation Segretation, Trennung
self-determination Selbstbestimmung
self-fulfilment Selbstverwirklichung
sewage Abwasser
sewer Abwasserkanal
shaft Schacht
shaft mining Schachtbergbau
share Anteil
to **share** teilen
shopping center Einkauszentrum
shoreline Küste
shredded zerkleinert
The vegetables were shredded.
to **shrink** schrumpfen
shrinkage Schrumpfen
sibling Geschwister
significance Bedeutung, Stellenwert, Wichtigkeit
significant bedeutend
to **simplify** vereinfachen
to **skew** verzerren, verdrehen
skill Können, Fähigkeit
skilled worker Facharbeiter
slaughterhouse Schlachthaus
smart grid intelligentes Stromnetz
social injustice soziale Ungleichheit
social security system Sozialversicherungssystem
soil compaction Bodenverdichtung
soil conservation Bodenschutz
soil degradation Bodendegradierung
soil erosion Bodenerosion
soil fertility Bodenfruchtbarkeit
solar cell Solarzelle
solar collector Sonnenkollektor
solar energy Sonnenenergie
solar panel Solarmodul
solar tower Solarturm
sound hier: gesund, einwandfrei
Sound sleep is good for people's health.
sovereignty Eigenständigkeit
special economic zone Sonderwirtschaftszone
specialisation Spezialisierung
specialised spezialisiert
specific bestimmt
sphere Bereich
to spread (sich) verbreiten
stability Stabilität, Beständigkeit
stable Stall
stage Phase, Stadium

stakeholder Akteur, Interessenvertreter
standard of living Lebensstandard
state funds staatliche Mittel
steady beständig
steel mill Stahlwerk
to **steer** lenken
stock Bestand
stock exchange Börse
storage facility Lager
storage reservoir Staubecken
to **strengthen** stärken
to **stress sth.** auf etw. besonderen Wert legen
striking auffällig
A striking performance by the football player
structural change Strukturwandel
structural data Stukturdaten
to **struggle with** kämpfen mit, sich abmühen mit
subscription hier: Vertrag
subsidy Zuschuss, Subvention
substitute Ersatz
suburb Vorort
suburbanisation Suburbanisierung
to **succeed** gelingen, glücken
successful erfolgreich
to **suffer from** leiden an
sufficient genügend, ausreichend
sugar beet Zuckerrübe
supplier Lieferant, Zulieferer
supply Versorgung
The water supply is unsafe.
supply chain Zulieferkette
to **support** unterstützen
supporter Unterstützer, Befürworter
supremacy Überlegenheit, Vormachtstellung
surface crusting Verkrustung, Krustenbildung
surplus Überschuss
to **sustain** stützen, erhalten
sustainability Nachhaltigkeit
sustainable development nachhaltige Entwicklung
Sustainable Development Goals (SDGs) Ziele für nachhaltige Entwicklung
sustainable forestry nachhaltige Forstwirtschaft
sustainable land use nachhaltige Landnutzung

T

tanker Tankschiff
to **target sb.** sich jdn. zur Zielgruppen nehmen
tariff wall Zollschranke
tax Steuer
tax cut Steuerersenkung
tax incentive Steuererleichterung
temporarily vorübergehend, zeitweise
tertiary sector tertiärer Sektor
the latter der/die/das Letztere
thereby dabei
therefore deshalb
thermal power station Wärmekraftwerk
to **thrive** Erfolg haben, gut gehen
New businesses thrive in this region.
tidal barrage Gezeitensperrwerk
tidal energy Gezeitenenergie
tidal flat Watt
tidal range Tidenhub
top soil Oberboden
to **trace back** zurückverfolgen
traffic Verkehr
traffic junction Verkehrsknotenpunkt
to **transfer** übertragen
to **transform** umsetzten, umwandeln
transformation Wandel, Umwandlung
transition Übergang
transnational company transnationales Unternehmen
to **trap sb.** einfangen
treatment Behandlung
treaty Abkommen
tree ring Baumring
triad Triade
to **trigger sth.** etw. auslösen, einleiten
trillion Billion
turbine Turbine
turnover Umsatz

U

ubiquitous allgegenwärtig
uncertainty Unsicherheit, Ungewissheit
unconventional oil Öl aus unkonventionellen Ressourcen
undernourished unterernährt
unemployment Arbeitslosigkeit
unemployment rate Arbeitslosenrate
unfavourable unvorteilhaft
unfortunately leider, unglücklicherweise
unintended unbeabsichtigt
unity Einheit
unprogressive rückständig
untapped unerschlossen
upswing Aufschwung
An upswing in the world economy
urban städtisch

V

to **vaccinate** impfen
valuable wertvoll
to **vary** sich verändern
vast weitläufig
vehicle Fahrzeug
versatile vielseitig
vessel Schiff
vicinity Umgebung
vicious circle Teufelskreis
victim Opfer
virtual water Virtuelles Wasser
visible sichbar, erkennbar
vital lebensnotwendig
voluntary service Freiwilligenarbeit

W

wage Lohn, Arbeitsentgelt
A weekly wage of €200
ward (Kranken-)Station
warehouse Lagerhalle
water availability Wasserverfügbarkeit
water distribution Wasserverteilung
water management Wasserwirtschaft
water scarcity Wasserknappheit
welfare system Sozialsystem
wheat Weizen
to **widen sth.** etw. erweitern, ausweiten
The gap between rich and poor widens.
wind energy Windkraft
wind farm Windpark
wind lens
wind turbine Windrad
to **withdraw** entnehmen, entziehen
to **withstand sth.** etw. standhalten
workforce Arbeitnehmerschaft
world economy Weltwirtschaft
to **worsen** verschlechtern

Y

yield Ertrag

Glossary

A

absorption [əb'sɔ:pʃn], p. 83
Absorption – A physical or chemical process in which atoms, molecules or ions enter some bulk phase – gas, liquid or solid material.

Agenda 21 [ə'dʒendə 'twentiwʌn], p. 103
Agenda 21 – A non-binding, voluntarily implemented action plan of the United Nations with regard to → sustainable development.

agglomeration [əˌglɒmə'reɪʃn], pp. 53, 111
Ballungsraum, Agglomeration – Major → urban area, often resulting from the merging of cities caused by the process of → suburbanisation.

agriculture ['ægrɪkʌltʃə(r)], p. 137
Landwirtschaft – The science or practice of → farming, concerning cultivating land, raising crops, feeding, breeding, and raising → livestock.

airline ['eəlaɪn], p. 145
Fluglinie – A company that provides regular flights to take passengers and goods to different places.

ambient heat ['æmbiənt hi:t], p. 75
Umgebungswärme – Heat energy available in the → environment, like in air, water, and soil.

ancillary industry [æn'sɪləri 'ɪndəstri], p. 119
Zulieferindustrie – Manufactures parts and components to be used in the production process; e.g. a company which produces engines needed for the production of a car.

assembly plant [ə'sembli pla:nt], p. 119
Montagewerk – Factory in which parts, e.g. of a car, are put together.

B

balance of trade [ˌbæləns əv 'treɪd], p. 125
Handelsbilanz – Difference in value between → imports and → exports.

biocapacity ['baɪəʊkə'pæsəti], p. 91
Biokapazität – The ability of the → environment to produce all the goods used by humans and to absorb the waste materials.

bioenergy [ˌbaɪəʊ'enədʒi], p. 69
Bioenergie – Energy contained in plants, animals, bacteria and fungi and their residues.

biofuel ['baɪəʊfju:əl], p. 69
Biokraftstoff – Gaseous, liquid, or solid substance composed of or produced from biological raw materials that is used as a → fuel.

biogas ['baɪəʊgæs], p. 69
Biogas – Gas that is produced by the action of bacterial decomposition of sewage, manure, garbage, or plant crops and used as a → fuel.

biomass ['baɪəʊmæs], p. 69
Biomasse – The total mass of all living things. In terms of energy production plant material and animal waste used as a fuel or energy source for the generation of electricity.

birth rate ['bɜ:θ reɪt], pp. 43, 45, 47
Geburtenrate – Figure, given in per mil (‰), describing the number of people born per thousand inhabitants in a certain area within one year.

bituminous coal [bɪ'tju:mɪnəs kəʊl], p. 59
Steinkohle – Usually black coal, used primarily as → fuel in steam-electric power generation, also used for heat and power applications in manufacturing.

brain drain ['breɪn dreɪn], p. 51
Brain Drain – Loss of highly skilled and qualified people to other countries where they hope to find more attractive jobs, i.e. to work in better conditions and to earn more money.

brain gain ['breɪn geɪn], p. 51
Brain Gain – → Immigration of highly trained or qualified people, regarded as beneficial to a country's → economy or society.

Brexit ['breksɪt], p. 119
Brexit – Merger of the words "Britain" and "exit" used to describe the British exit from the → EU.

business district ['bɪznəs 'dɪstrɪkt], p. 151
Geschäftsviertel – Part of a city or town where there are many businesses.

C

capital ['kæpɪtl], p. 151
Hauptstadt – City with a seat of government.

car manufacturer [ka:(r) ˌmænju'fæktʃərə(r)], p. 119
Autohersteller – Company which produces cars.

carbon cycle ['ka:bən saɪkl], p. 83
Kohlenstoffkreislauf – The process by which carbon is exchanged among the spheres (pedosphere, geosphere, hydrosphere, and atmosphere) of the Earth.

carbon emission trading ['ka:bən i'mɪʃn 'treɪdɪŋ], p. 83
Emissionshandel mit Kohlendioxid – A government-mandated instrument to reduce → greenhouse gas → emissions.

carbon footprint [ˌka:bən 'fʊtprɪnt], p. 83
CO_2-Fußabdruck – A measure of the total amount of carbon dioxide (CO_2) and methane (CH_4) emissions of a defined population, system or activity, considering all relevant sources, sinks and storage within the spatial and temporal boundary of the population, system or activity of interest.

Central Business District (CBD) ['sentrəl 'bɪznəs 'dɪstrɪkt], p. 31
Hauptgeschäftszentrum – Area of a town or a city which is the centre of administration, shopping, business and entertainment.

circular economy ['sɜ:kjələ(r) ɪ'kɒnəmi], p. 99
Kreislaufwirtschaft – Economic system in which resource input and waste, → emission, and energy leakage are minimised through long-lasting design, sharing, repair, reuse, refurbishing, remanufacturing, and recycling.

clean coal technology [kli:n kəʊl tek'nɒlədʒi], p. 59
saubere Kohletechnologie – Reduces harmful environmental effects like CO_2 and particulates → emissions by using multiple filter, conversion, and storage technologies.

climate agreement ['klaɪmət ə'gri:mənt], p. 81
Klimaabkommen – Binding agreement between nations of the world on tackling → climate change by cutting → greenhouse gas → emissions.

climate change ['klaɪmət tʃeɪndʒ], p. 81
Klimawandel – Change in global or regional climates over a long period of time; the variation of climatic elements over time.

climate protection [ˈklaɪmət prəˈtekʃn], p. 81
Klimaschutz – An agreement to protect the global climate; mainly the stabilization of → greenhouse gas concentrations in the atmosphere at a level that would prevent dangerous anthropogenic interference with the climate system.

coal mining crisis [kəʊl ˈmaɪnɪŋ ˈkraɪsɪs], pp. 141, 143
Kohlenkrise – Crisis of the coal mining industry that started in the 1950s and hit the → old industrial regions of Europe, for example. Its consequences include jobloss and the decline of industrial regions due to a sinking demand for coal and falling prices.

combined heat and power (CHP) [kəmˈbaɪnd hiːt ənd ˈpaʊə(r)], p. 59
Kraft-Wärme-Kopplung – Combines the production of usable heat and electricity in one single, efficient process.

commodity [kəˈmɒdəti], p. 61
Rohstoff, Handelsgut – Basic goods used in commerce that is often used as input in the production of other goods.

Common Agricultural Policy (CAP) [ˈkɒmən ˌægrɪˈkʌltʃərəl ˈpɒləsi], p. 115
Gemeinsame Agrarpolitik (GAP) – Agricultural policy of the → EU.

component supplier [kəmˈpəʊnənt səˈplaɪə(r)], p. 17
Zulieferer / Zulieferfirma – Company producing parts (components) for the final assembly in another (larger and less-specialised) company.

concentrated solar power (CSP) [ˈkɒnsntreɪtɪd ˈsəʊlə(r) ˈpaʊə(r)], p. 73
solarthermische Energie – Mirrors or lenses are used to focus a large area of sunlight onto a small area.

container vessel [kənˈteɪnə(r) ˈvesl], p. 13
Containerschiff – Type of ship that is build to be used in global trade to ship containers.

contamination [kənˌtæmɪˈneɪʃn], p. 85
Verschmutzung – Presence of unwanted substances in the natural → environment.

conventional gas [kənˈvenʃənl gæs], p. 63
Gas aus konventionellen Ressourcen – → Natural gas occurring in porous and permeable reservoir rock, either in the gaseous phase or dissolved in → crude oil, which technically can be produced by normal production.

crude oil [kruːd ɔɪl], p. 61
Rohöl – Petroleum as it comes from the ground, before refining.

D

death rate [ˈdeθ reɪt], pp. 43, 45, 47
Sterberate – Figure, given in per mil (‰), describing the number of deaths per thousand inhabitants in a certain area within one year.

decarbonisation [ˌdiːˌkɑːbənaɪˈzeɪʃn], p. 77
Entkarbonisierung – Reduction of the use of → fossil fuels which all produce carbon dioxide.

decentralisation [ˌdiːˌsentrəlaɪˈzeɪʃn], p. 77
Dezentralisierung – Dispersion of a concentration of political power, industry, or population.

deforestation [ˌdiːˌfɒrɪˈsteɪʃn], p. 33
Abholzung, Entwaldung – Clearing of forest to provide firewood, building materials or to use the land for economic activities.

demographic change [ˌdeməˈgræfɪk tʃeɪndʒ], p. 127
demographischer Wandel – Shift in population structure that has emerged in many industrialised nations since the 1970s, chiefly as a result of three major developments: declining → birth rates, longer → life expectancy and increasing → immigration.

demographic transition [ˌdeməˈgræfɪk trænˈzɪʃn], p. 101
demographischer Übergang – Economic theory that links population changes to levels of economic, education, and healthcare development. It states that as women become better educated and financially independent, the global → fertility rates will continue to decline. Low → birth rates combined with low → death rates will result in an increasing number of older people dependent on pension schemes.

Demographic Transition Model (DTM) [ˌdeməˈgræfɪk trænˈzɪʃn ˈmɒdl], p. 45
Modell des demographischen Übergangs – A model that describes how → birth and death rates shift from high to low levels over time as countries' economies develop and living conditions improve. The decline in mortality usually precedes the decline in fertility, thus resulting in rapid → population growth during the transition period.

dependent [dɪˈpendənt], p. 47
Abhängige(r), hier: nicht Erwerbstätige(r) – Non-working population below 15 and above 65 years of age.

deposit [dɪˈpɒzɪt], p. 63
Lagerstätte – Part of the Earth's crust where solid, liquid, or gaseous → resources are found in a high natural concentration.

development [dɪˈveləpmənt], p. 37
Entwicklung – Examination of the state of economic richness, standard of living, quality of life and the state of the → environment (development geography).

digital divide [ˈdɪdʒɪtl dɪˈvaɪd], p. 15
digitale Kluft – Differences in accessibility and usage of information and communication technology in different regions of the Earth.

disparity [dɪˈspærəti], p. 25
Unterschied, Disparität – A difference, especially one connected with unfair treatment.

district heating [ˈdɪstrɪkt ˈhiːtɪŋ], p. 59
Fernwärme – System for distributing heat generated in a centralized location for residential and commercial space heating and water heating.

downtown [ˌdaʊnˈtaʊn], p. 31
Innenstadt, Stadtzentrum – Central part of a city, especially its main commercial area.

Dust Bowl [ˈdʌst bəʊl], p. 97
Dust Bowl – Area of land in the central USA that has been turned into desert by lack of rain or too much → farming.

E

ecological footprint [ˌiːkəˈlɒdʒɪkl ˈfʊtprɪnt], p. 91
Ökologischer Fußabdruck – Way to measure the human demands on Earth to produce all the goods and services needed and to assimilate the waste.

ecological overshoot [ˌiːkəˈlɒdʒɪkl ˌəʊvəˈʃuːt], p. 91
Überschreitung der ökologischen Belastbarkeit – Situation in which human activities exceed the biocapacity.

economic crisis [ˌi:kəˈnɒmɪk ˈkraɪsɪs], p. 31
Wirtschaftskrise – A situation in which the → economy of a country experiences a sudden downturn brought on by a financial crisis. An economy facing an economic crisis will most likely experience a falling → GDP, a drying up of liquidity and rising / falling prices due to inflation / deflation.

economic power [ˌi:kəˈnɒmɪk ˈpaʊə(r)], p. 109
Wirtschaftskraft – The amount of productive resources that are available and that give, e.g., a country the ability to make and enforce economic decisions.

economic sector [ˌi:kəˈnɒmɪk ˈsektə(r)], p. 125
Wirtschaftssektor – Category of activities in the production of goods, e.g. a division of a country's population based upon the economic area in which that population is employed.

economy [ɪˈkɒnəmi], p. 11
Wirtschaft – Sum of all activities of the → primary, → secondary, and → tertiary sector (industries).

electric mobility [ɪˈlektrɪk məʊˈbɪləti], p. 135
Elektromobilität – Concept of → Energiewende; all vehicles that primarily use energy taken from the electric energy.

emigration [ˌemɪˈgreɪʃn], pp. 51, 133
Auswanderung – Movement of people out of a country.

emission [iˈmɪʃn], p. 83
Emission, Ausstoß – Sending out of substances.

Energiewende [], p. 135
Energiewende – Part of a German politics which aims to reduce the use of fossil → fuels and abolish nuclear → fuels, creating an energy system based on at least 60% → renewable energy by 2050.

energy conversion [ˈenədʒi kənˈvɜ:ʃn], p. 69
Energieumwandlung – Transformation of one form of energy to another, always connected with an energy loss.

energy crop [ˈenədʒi krɒp], p. 69
Energiepflanze – Food crop and non-food crop grown for its fuel value.

energy efficiency [ˈenədʒi ɪˈfɪʃnsi], pp. 77, 135
Energieeffizienz – Percentage of total energy input that is useful and not wasted as useless heat.

energy transition [ˈenədʒi trænˈzɪʃn], p. 135
Energiewende – Transition towards sustainable energy, decentralised → renewable energy, and → energy efficiency.

environment [ɪnˈvaɪrənmənt], pp. 39, 85, 89, 115, 139
Umwelt – All living and non-living things with their interactions that affect an organism or a group of organisms.

erosion [ɪˈrəʊʒn], p. 33
Bodenabtrag, Erosion – Destructive effect on the Earth's surface caused by running water, wind, moving ice, and gravity, leading to the transport of material.

ethnic group [ˈeθnɪk gru:p], p. 27
ethnische Gruppe – Group of people who identify with each other based on common social, cultural or national experience.

euro [ˈjʊərəʊ], pp. 107, 109
Euro (Währung) – Most members of the → European Union have introduced a common currency, the euro.

European Central Bank (ECB) [ˌjʊərəˈpi:ən ˌsentrəl ˈbæŋk], p. 145
Europäische Zentralbank (EZB) – Bank responsible for the monetary system of the → European Union and the Euro currency.

European Community (EC) [ˌjʊərəˈpi:ən kəˈmju:nəti], p. 109
Europäische Gemeinschaft (EG) – Used to be the European Economic Community (EEC); an organisation created to integrate the European economies.

European Union (EU) [ˌjʊərəˈpi:ən ˈju:niən], pp. 107, 109, 121
Europäische Union (EU)– Economic and political organisation that many European countries belong to.

Euroskepticism [ˈjʊərəʊ ˈskeptɪsɪzəm], p. 121
EU-Skepsis – Very critical view of the → EU.

export [ɪkˈspɔ:t], p. 21
Ausfuhr (Export) – Process of selling goods or services to another country as well as their financial value.

export good [ɪkˈspɔ:t gʊd], p. 125
Exportgut – Product that is sold and sent from one country to another country.

extraction [ɪkˈstrækʃn], p. 63
Förderung – Process by which → crude oil or → natural gas is drawn out from beneath the earth's surface.

F

farming [ˈfɑ:mɪŋ], p. 97
Landwirtschaft – The use of land for growing crops or breeding livestock.

fish farm [ˈfɪʃ fɑ:m], p. 89
Fischzuchtbetrieb – A form of aquaculture to raise fish commercially in tanks or enclosures, usually for food.

fish stock [ˈfɪʃ stɒks], p. 89
Fischbestand – A subpopulation of a particular species of fish.

fishery [ˈfɪʃəri], p. 33
Fischerei – The occupation or industry of catching or rearing fish.

fission [ˈfɪʃn], p. 65
Kernspaltung – A process in which the nucleus of a heavy atom is split apart and energy released.

flood protection [flʌd prəˈtekʃn], p. 87
Hochwasserschutz – All methods used to reduce or prevent the effects of flood waters or high water levels.

food security [fu:d sɪˈkjʊərəti], p. 49
Ernährungssicherheit – Everybody has access to enough nutritious, safe and affordable food to live a healthy and active life.

Fordism [fɔ:dɪzəm], p. 31
Fordismus – The use in the manufacturing industry of the methods pioneered by Henry Ford, typified by large-scale, mechanised mass production.

foreign direct investment (FDI) [ˈfɒrən dəˈrekt ɪnˈvestmənt], p. 35
ausländische Direktinvestition – An investment in a business by an investor from another country who controls ownership (e.g. real estate, companies, business shares).

forestry [ˈfɒrɪstri], p. 137
Forstwirtschaft – The cultivation of trees.

fossil fuel [ˈfɒsl fju:əl], pp. 57, 59
fossiler Brennstoff – A → fuel (coal, oil, or → natural gas) formed in the earth from plant or animal remains.

fracking [ˈfrækɪŋ], p. 63
Fracking – (hydraulic fracturing) A method of extracting oil or gas from a layer of rock by drilling and creating cracks.

fragmentation [ˌfrægmenˈteɪʃn], pp. 25, 39
Fragmentierung, Zersplitterung – Increasing social or economic inequalities in the process of → globalisation.

free-market economy [ˌfriː ˈmɑːkɪt ɪˈkɒnəmi], p. 35
freie Marktwirtschaft – Economic system with a government not interfering in business activities.

free trade zone [ˌfriː treɪd zəʊn], pp. 25, 35
Freihandelszone – A special area within a country where foreign companies can import materials, manufacture goods, export products, etc. without being limited by the usual rules and taxes.

freight [freɪt], p. 13
Fracht / Frachtgut – Goods that are transported for financial gain.

fresh water [freʃ ˈwɔːtə(r)], p. 87
Trinkwasser – The naturally occurring water on Earth's surface in ice sheets, ice caps, glaciers, icebergs, bogs, ponds, lakes, rivers and streams, and groundwater in aquifers and underground streams.

fuel [fjuːəl], p. 61
Brennstoff – An energy source for engines, → power stations, or reactors.

fuel cell [fjuːəl sel], p. 77
Brennstoffzelle – Device that produces electricity through a chemical reaction of a source fuel directly from the oxidation of a → fuel, as that of hydrogen by oxygen.

G

general fertility rate [ˈdʒenrəl fəˈtɪləti reɪt], p. 43
Fruchtbarkeitsrate – Figure, usually given in per mil (‰), showing the average number of children born to a woman in her lifetime.

gentrification [ˌdʒentrɪfɪˈkeɪʃn], pp. 147, 151
Gentrifizierung – Movement of wealthy people mostly into the inner city after the renovation and restoration of buildings and flats; mostly combined with the outward → migration of the former inhabitants who cannot afford living there any longer.

geothermal energy [ˌdʒiːəʊˈθɜːml ˈenədʒi], p. 75
Geothermalenergie – Energy produced by extracting the earth's internal heat, originating in its core.

global city [ˈgləʊbl ˈsɪti], p. 27
globale Stadt – Cities which are the most important political, economic, cultural and financial centres of the global economic system.

global player [ˈgləʊbl ˈpleɪə(r)], p. 17
Global Player, Weltkonzern – Corporation or enterprise that manages production or delivers services on a worldwide scale (multinational corporation).

global sourcing [ˈgləʊbl sɔːsɪŋ], p. 17
weltweite Beschaffung – Economic strategy with the aim of finding the most cost efficient location for manufacturing a product, even if the location is in a foreign country.

global warming [ˌgləʊbl ˈwɔːmɪŋ], pp. 59, 81, 83
Erderwärmung – An increase in the earth's average atmospheric temperature that causes corresponding changes in climate and that likely results from the → greenhouse effect.

globalisation [ˌgləʊbəlaɪˈzeɪʃn], pp. 11, 39
Globalisierung – Process of increasing the interconnection of different national and regional economies, societies, and cultures. Its striking elements are global networks of trade, communication, transportation, and → migration.

government district [ˌgʌvnˈment ˈdɪstrɪkt], p. 151
Regierungsviertel – An administrative region at the federal state level in Germany.

government institution [ˌgʌvnˈment ˌɪnstɪˈtjuːʃn], p. 151
staatliche Institution – An established organization or foundation, especially one dedicated to education, public service, or culture run by a government.

greenhouse effect [ˈgriːnhaʊs ɪfekt], p. 81
Glashaus-/Treibhauseffekt – Warming of the atmosphere through gases like carbon dioxide, methane and water vapour which allow short-wave radiation (light) to pass through the atmosphere but absorb some of the long-wave radiation (heat) emitted from the Earth's surface.

greenhouse gas [ˌgriːnhaʊs ˈgæs], pp. 59, 81
Treibhausgas – A gas in an atmosphere that absorbs and emits radiation. Its absorption of solar radiation is responsible for the → greenhouse effect, e.g. carbon dioxide, methane, ozone, and water vapour.

grid expansion [grɪd ɪkˈspænʃn], p. 135
Netzausbau – Building a network of high voltage cables combined with a → smart grid in order to connect (renewable) energy production sites with consumers.

gross domestic product (GDP) [ˌgrəʊs dəˌmestɪk ˈprɒdʌkt], p. 25
Bruttoinlandsprodukt (BIP) – Value of all finished goods and services which are produced or rendered in a country within one year.

gross national product (GNP) [ˌgrəʊs ˈnæʃnəl ˈprɒdʌkt], p. 37
Bruttonationalprodukt (BNP) – The total value of all the goods and services produced by a country in one year, including the total income from foreign countries.

H

headquarters [ˌhedˈkwɔːtəz], p. 147
Hauptsitz, Hauptniederlassung – The headquarters of a company are its main offices.

health care [ˈhelθ keə(r)], p. 47
Gesundheitsversorgung – The service of providing medical care.

heat pump [hiːt pʌmp], p. 75
Wärmepumpe – Device that takes heat from one location and moves it to another location either to heat or cool.

heavy metal [ˌhevi ˈmetl], p. 85
Schwermetall – Metal with relatively high density, atomic weight, or atomic number.

high-quality good [haɪ ˈkwɒləti gʊd], p. 125
hochwertige Ware – A product of superior quality.

high-tech industry [ˌhaɪ ˈtek ˈɪndəstri], p. 29
High-Tech-Industrie – Industry using advanced skills and machines, especially electronic ones.

hub [hʌb], pp. 29, 145
Drehscheibe, Knotenpunkt – The effective centre of an activity, region, or network.

hub-and-spokes system [hʌb ənd spəʊks ˈsɪstəm], p. 13
Speichennetz – Air transport using a limited number of central airports where passengers or goods are transferred from one plane to another to reach the final destination.

Human Development Index (HDI) [ˈhjuːmən dɪˈveləpmənt ˈɪndeks], p. 37
Index der menschlichen Entwicklung – Number which helps to compare the degree of human development in countries.

It is based on life expectancy, literacy rate, the educational system, and → GDP per capita.

hunger [ˈhʌŋɡə(r)], p. 49
Hunger – The state of not having enough food to eat, especially when this causes illness or death.

hydroelectric dam [ˌhaɪdrəʊɪˈlektrɪk dæm], p. 67
Wasserkraft-Talsperre – A dam containing generators to make electricity.

hydropower [ˈhaɪdrəʊˈpaʊə(r)], p. 67
Wasserkraft – Electrical energy gained by running water.

I

immigration [ˌɪmɪˈɡreɪʃn], pp. 51, 127, 131
Einwanderung – The movement of people into a country.

import [ˈɪmpɔːt], p. 21
Einfuhr (Import) – Process of buying goods or services from another country as well as their financial value.

import good [ˈɪmpɔːt ɡʊd], p. 125
Importgut – A product that is brought into one country from another.

industrial agriculture [ɪnˈdʌstriəl ˈæɡrɪkʌltʃə(r)], p. 137
industrielle Landwirtschaft – A form of modern → farming that refers to the industrialised production of → livestock, → poultry, fish, and crops. It includes e.g. innovation in agricultural machinery and farming methods, and genetic technology.

industrial park [ɪnˈdʌstriəl pɑːk], p. 145
Industriegebiet – An area especially for factories, on the edge of a town.

infant mortality rate [ˈɪnfənt mɔːˈtæləti reɪt], p. 45
Säuglingssterblichkeitsrate – Figure, usually given in per mil (‰), showing the number of deaths of children under age 1 within one year.

informal economy [ɪnˈfɔːml ɪˈkɒnəmi], p. 97
Schattenwirtschaft – The production of goods or services to generate, mostly casual, employment and incomes outside the supervision of the state.

informal sector [ɪnˈfɔːml ˈsektə(r)], p. 97
informeller Sektor – Sector which encompasses all jobs which are not recognized as normal income sources, and on which taxes are not paid.

infrastructure [ˈɪnfrəstrʌktʃə(r)], pp. 19, 47, 117, 119, 131, 149
Infrastruktur – Facilities which provide the services for economic and individual life; e.g. railway lines, energy and water supply, communication systems, schools, hospitals.

in-migration [ɪn maɪˈɡreɪʃn], p. 131
Zuwanderung – The movement of people into a new area from elsewhere in the same country.

in-situ leaching (ISL) [ˌɪn ˈsɪtjuː liːtʃɪŋ], p. 65
Untertagelaugung – Leaves the ore where it is, in the ground, and recovers the minerals from it by dissolving them with a solution and pumping the mineral-enriched solution to the surface where the minerals can be recovered.

internet [ˈɪntənet], p. 15
Internet – The global system of interconnected computer networks that use the internet protocol suite (TCP/IP) to link devices worldwide.

J

joint venture [ˌdʒɔɪnt ˈventʃə(r)], p. 35
Zusammenschluss von Unternehmen – Business activity that involves at least two companies doing something together.

just-in-time production [dʒʌst ɪn taɪm prəˈdʌkʃn], p. 17
Produktion auf Abruf, Just-in-time Produktion – Production of the necessary items in necessary quantities at the necessary time.

L

labour [ˈleɪbə(r)], p. 11
Arbeit – One of the three factors of production. Labour is the work done by people usually in order to earn money.

labour force [ˈleɪbə(r) fɔːs], p. 19
Arbeiterschaft – Term for either all the people working for a certain company, or all the potential workers in a certain area or region.

lean production [liːn prəˈdʌkʃn], p. 17
schlanke Produktion – Form of production in which costs are cut by cutting out all activities that do not add value to the production process as well as by outsourcing production steps to specialised companies or enterprises.

Less Economically Developed Country (LEDC) [les ˌiːkəˈnɒmɪkli dɪˈveləpt ˈkʌntri], pp. 15, 37, 43
Entwicklungsland – Country where → GDP per capita, literacy rate, food supply, and → life expectancy are low, whereas → population growth, infant mortality, and the share of the agricultural sector are high. The poorest of these countries are called least economically developed countries (LDC).

life expectancy [ˈlaɪf ɪkspektənsi], pp. 37, 45
Lebenserwartung – Figure, given in years, describing the average age a newborn can expect to reach in that country.

lignite [ˈlɪɡnaɪt], p. 59
Braunkohle – A soft, brown coal, with a low energy content often having a distinct wood-like texture.

liquefied natural gas (LNG) [ˈlɪkwɪfaɪd ˌnætʃrəl ˈɡæs], p. 63
verflüssigtes Erdgas – → Natural gas compressed at moderate pressure and cooled to -161 °C to remain liquid and then stored and transported in insulated pressure containers.

livestock farming [ˈlaɪvstɒk ˈfɑːmɪŋ], p. 139
Tierhaltung, Viehwirtschaft – Rearing of farm animals for the production of milk or meat.

living standard [ˈlɪvɪŋ ˈstændəd], p. 111
Lebensstandard – A term used to refer to a level of comfort that people have. It depends on several factors, such as wealth, available necessities, economic and political stability, political and religious freedom, environmental quality, climate, and safety. It is closely related to → quality of living.

Local Agenda 21 [ˈləʊkl əˈdʒendə ˈtwentiwʌn], p. 103
Regionale Agenda 21 – The implementation of → Agenda 21 on a local level by local authorities.

location factor [ləʊˈkeɪʃn ˈfæktə(r)], pp. 19, 29
Standortfaktor – Factors, which affect an enterprise in the choice of its site. There are hard location factors (e.g. availability of raw materials, transportation network, taxes, etc.) and soft location factors (e.g. cultural offer, leisure facilities, educational provision, etc.).

M

malnutrition [ˌmælnjuːˈtrɪʃn], p. 49
Fehlernährung – Malnutrition results from deficiencies, excesses or imbalances in the consumption of nutrients.

manufacturer [ˌmænjuˈfæktʃərə(r)], p. 119
Hersteller, Produzent – A company producing goods in factories.

marine energy [məˈriːn ˈenədʒi], p. 75
Meeresenergie – All types of energy present in the ocean or along the coast generated by solar energy, gravity, temperature or salt gradient.

marine pollution [məˈriːn pəˈluːʃn], p. 89
Meeres-/Ozeanverschmutzung – The contamination of the oceans mainly by human activities.

market [mɑːkɪt], p. 11
Markt – A physical place where goods are traded as well as the system of trading goods and services in exchange for money.

megacity [ˈmegəsɪti], p. 53
Megacity – A metropolitan area with a total population of more than 10 million people.

methane [ˈmiːθeɪn], p. 63
Methan – A potent → greenhouse gas which is a main component of → natural gas and a product of some biological breakdown processes without oxygen.

metropolitan area [ˌmetrəˈpɒlɪtən ˈeəriə], p. 117
Ballungsraum – A very large urban settlement consisting of cities surrounded by urban areas with a total population exceeding 1,000,000 inhabitants.

micro-hydro power station [ˈmaɪkrəʊ ˈhaɪdrəʊ ˈpaʊə steɪʃn], p. 67
Mini-Wasserkraftwerk – In a "run-of-the-river" system water is diverted from the stream or river and is redirected back into the same watercourse, driving a → turbine which produces up to 100 kW.

migration [maɪˈgreɪʃn], pp. 51, 53, 131
Wanderung – The movement of people (migrants) or animals from one place to another.

migration policy [maɪˈgreɪʃn ˈpɒləsi], p. 127
Migrationspolitik – National policy approaches for designing and managing → in- and out-migration, in particular legal labour migration, as well as data collection and assessment.

modernisation [ˌmɒdənaɪˈzeɪʃn], p. 139
Modernisierung – The process of adapting something to modern needs or habits.

More Economically Developed Country (MEDC) [mɔː(r) ˌiːkəˈnɒmɪkli dɪˈveləpt ˈkʌntri], pp. 15, 37, 43
wirtschaftlich stärker entwickeltes Land – Country that characteristically shows high → GDP per capita, literacy rate, food supply, and → life expectancy, as well as low → population growth, infant mortality, and share of the agricultural sector.

multinational company [ˌmʌltiˈnæʃnəl ˈkʌmpəni], p. 39
multinationaler Konzern – A company operating in several countries, but managed from one (home) country.

multinational corporation [ˌmʌltiˈnæʃnəl ˌkɔːpəˈreɪʃn], p. 21
multinationale Gesellschaft – A worldwide enterprise that owns and produces goods or services in one or more countries apart from their home country.

N

national economy [ˈnæʃnəl ɪˈkɒnəmi], p. 125
Volkswirtschaft – A national economy refers to the economy of an entire country, including its financial resources and management. It encompasses the value of all goods and services manufactured within a nation.

natural gas [ˌnætʃrəl ˈgæs], p. 63
Erdgas – A mixture of gases (→ methane, nitrogen, carbon dioxide etc.) located near coal and → crude oil beds.

natural resource [ˈnætʃrəl rɪˈsɔːs], p. 19
natürliche Lebensgrundlage – Any property of the physical environment, such as minerals, or natural vegetation, which humans can use to satisfy their needs, especially in order to produce goods. Natural resources may be classified as renewable or non-renewable resources.

net migration gain [net maɪˈgreɪʃn geɪn], p. 133
Nettozuwanderung – Result of more people immigrating to, than emigrating from, a region.

net migration loss [net maɪˈgreɪʃn lɒs], p. 133
Nettoabwanderung – Result of more people emigrating from, than immigrating to, a region.

Newly Industrialized Country (NIC) [ˈnjuːli ɪnˈdʌstriəlaɪzd ˈkʌntri], p. 37
Schwellenland – Country whose national → economy has transitioned from being primarily based in agriculture to being primarily based in goods-producing industries (e.g. manufacturing). It has a higher standard of living than → LEDCs, but has not yet reached the level of economic advancement of → MEDCs.

non-conventional gas [nɒn kənˈvenʃənl ˈgæs], p. 63
Gas aus unkonventionellen Ressourcen – A → natural gas found in unusual underground situations such as very impermeable reservoirs, hydrates, and coal → deposits.

non-renewable energy [nɒn rɪˈnjuːəbl ˈenədʒi], p. 57
nichterneuerbare Energie – A natural resource such as a → fossil → fuel or uranium, which, once consumed, cannot be replaced.

nuclear disaster [ˈnjuːkliə(r) dɪˈzɑːstə(r)], p. 65
Atomunfall – Incident that results in the release of radioactive material from a nuclear facility.

nuclear energy [ˈnjuːkliə(r) ˈenədʒi], p. 65
Kernenergie – Energy that is created by splitting apart or fusing the nuclei of atoms.

nuclear fuel cycle [ˈnjuːkliə(r) ˈfjuːəl ˈsaɪkl], p. 65
Kernbrennstoffkreislauf – Radioactive uranium, after being used to produce electricity, needs storing or reprocessing and recycling to be used again in another cycle.

nuclear power station [ˈnjuːkliə(r) ˈpaʊə steɪʃn], p. 65
Kernkraftwerk – A factory where → nuclear energy is converted into heat, electricity, etc.

O

offshore [ˌɒfˈʃɔː(r)], p. 71
auf offener See – At sea.

oil spill [ɔɪl spɪl], p. 61
Ölverschmutzung – An accident in which oil has been lost during transport and caused pollution.

oil well [ˈɔɪl wel], p. 61
Ölbohrloch – A hole drilled into the earth or sea bed for the extraction of petroleum.

old industrial region [əʊld ɪn'dʌstriəl 'ri:dʒən], pp. 141, 143
Altindustrieregion – Industrial region whose rise was usually based on coal mining and the steel industry. Most of these regions have undergone some form of rapid decline in the second half of the 20th century and may be in the process of structural change.

onshore ['ɒnʃɔ:(r)], p. 71
an Land – On land.

OPEC ['əʊpek], p. 61
OPEC – Organisation of oil exporting countries with 13 member states (2017)

organic farming [ɔ:'gænɪk 'fɑ:mɪŋ], p. 115
biologische Landwirtschaft – Production of food and other agricultural products on the basis of principles that are environmentally friendly and allow for farm animal welfare and housing.

out-migration [aʊt maɪ'greɪʃn], p. 131
Abwanderung – The movement of people outward from one place in a country to another one in the same country.

outsourcing ['aʊtsɔ:sɪŋ], pp. 17, 25
Outsourcing, Auslagerung – Transfer of tasks, services or the manufacturing of products to another company. Typically used by companies to save costs.

overfishing [ˌəʊvə'fɪʃɪŋ], pp. 33, 89
Überfischung – Fishing at a rate that exceeds the rate which is necessary to reproduce the → fish stock.

overnutrition ['əʊvənju'trɪʃn], p. 49
Überernährung – People suffer from overnutrition if they consume too many calories.

P

passenger ['pæsɪndʒə(r)], p. 13
Passagier – A person who is travelling in a car, bus, train, plane or ship and who is not driving it or working on it.

petrochemical industry [ˌpetrəʊ'kemɪkl 'ɪndəstri], p. 61
Petrochemie – A heavy industry producing synthetic materials, mainly from products derived from the refining of petroleum and → natural gases.

petrol ['petrəl], p. 61
Benzin – A liquid refined from petroleum, used especially as a → fuel for cars and other vehicles.

photovoltaic (PV) ['fəʊtəʊvɒltɑ:ɪk], p. 73
Photovoltaik – Direct conversion of solar energy into electrical energy through → solar cells.

pipeline ['paɪplaɪn], p. 63
Pipeline – A line of pipes with pumps, valves, and control devices to transport liquids, gases, or finely dispersed solids.

plastic ['plæstɪk], p. 89
Plastik, Kunststoff – A material consisting of a wide range of synthetic or semi-synthetic organic compounds. They are usually synthetic, most commonly derived from petrochemicals, but many are made from renewable materials.

point-to-point system [pɔɪnt tə pɔɪnt 'sɪstəm], p. 13
Punkt-zu-Punkt System – A transportation system (air, rail, road, water) linking destinations directly.

pollution [pə'lu:ʃn], p. 33
Verschmutzung – Contamination of the → environment or parts of it (soil, air, water) caused by human activity.

population distribution [ˌpɒpju'leɪʃn ˌdɪstrɪ'bju:ʃn], p. 43
Bevölkerungsverteilung – The spread of people over an area.

population growth [ˌpɒpju'leɪʃn grəʊθ], pp. 43, 101
Bevölkerungswachstum – An increase in the number of people that reside in a country, state, county, or city.

population growth rate [ˌpɒpju'leɪʃn grəʊθ reɪt], pp. 43, 45
Bevölkerungswachstumsrate – Figure, given as a percentage (%) or in per mil (‰), describing the number of people added to (or subtracted from) a population of a given area in a year due to natural increase or decrease and → migration.

population structure [ˌpɒpju'leɪʃn 'strʌktʃə(r)], p. 47
Bevölkerungsstruktur – The age and sex composition of a country, region or city.

poultry farming ['pəʊltri 'fɑ:mɪŋ], p. 139
Geflügelzucht, -Haltung – Raising of birds domestically or commercially, primarily for meat and eggs.

poverty ['pɒvəti], pp. 39, 47
Armut – Minimum level of income that is necessary for a person to meet his/her basic needs. What is necessary to satisfy basic needs, differs from country to country. It depends on the country's level of development, norms and values. If a person's income is below this level, he or she is said to live in poverty.

power station ['paʊə steɪʃn], p. 59
Kraftwerk – A factory where electricity is generated.

power-to-gas (P2G) ['paʊə tə 'gæs], p. 135
Power-to-Gas – Technologies that use surplus electrical power, mostly from renewable resources, to make hydrogen or methane from water and CO_2.

precision farming [prɪ'sɪʒn 'fɑ:mɪŋ], p. 137
Präzisionslandwirtschaft – Modern → farming practice managing crop production inputs (seed, fertilizer, pesticides, etc.) on a site-specific basis to increase profits, reduce waste and maintain environmental quality.

primary energy ['praɪməri 'enədʒi], p. 57
Primärenergie – Any → renewable or → non-renewable energy that has not been subjected to any conversion or transformation process.

primary sector ['praɪməri 'sektə(r)], p. 125
primärer Sektor – The sector of an economy making direct use of natural resources or exploiting → natural resources. This includes agriculture, forestry, fishing and mining.

production value [prə'dʌkʃn 'vælju:], p. 137
Produktionswert – Turnover or revenue from sales of goods and rendering of services.

pull factor [pʊl 'fæktə(r)], p. 51
Pull Faktor – Social, political, economic or environmental conditions which attract people to a new location.

pumped-storage power station [pʌmpt 'stɔ:rɪdʒ 'paʊə steɪʃn], p. 67
Pumpwasserkraftwerk – A → power station using power at a period of low demand to pump water back up to a high storage reservoir so that it can be released to generate electricity at a period of high demand.

push factor [pʊʃ 'fæktə(r)], p. 51
Push Faktor – Social, political, economic or environmental conditions which make people leave their old location.

Q

quality of life ['kwɒləti əv 'lɪve], p. 147
Lebensqualität – Factors describing the positive living conditions within a society for its individuals. The term is mainly used as a measure of well-being of societies. Therefore, it can also be regarded as a very subjective measure of happiness.

R

radioactive waste [ˌreɪdiəʊ'æktɪv weɪst], p. 65
Atommüll – The radioactive by-products from the operation of a nuclear reactor or from depleted radioactive material.

ragpicker [ræg'pɪkə(r)], p. 99
Lumpensammler – A person who makes a living by picking up rags and other waste material from the streets, dumps, etc.

rate of natural increase [reɪt əv 'nætʃrəl ɪn'kri:s], pp. 43, 127
natürliche Wachstumsrate – The crude → birth rate minus the crude → death rate, excluding the population increase from → in-migration and → out-migration.

raw material [ˌrɑ: mə'tɪəriəl], p. 21
Rohstoff – Unprocessed material used for the production of goods.

recycling [ˌri:'saɪklɪŋ], p. 99
Recycling – The process of treating things that have already been used so that they can be used again.

refinery [rɪ'faɪnəri], p. 61
Raffinerie – A factory where a substance such as oil is made pure (= refined).

regional disparity ['ri:dʒənl dɪ'spærəti], pp. 111, 113, 129
regionale Ungleichheit – Difference between areas or regions in terms of population density, economic power, natural conditions, etc.

regional policy ['ri:dʒənl 'pɒləsi], p. 113
Regionalpolitik – A policy concerned with promoting a balanced distribution of industrial activity, employment and wealth between the various geographical areas of a country.

reindustrialisation [ri:ɪnˌdʌstriəlaɪ'zeɪʃn], p. 149
Reindustrialisierung – The process of organising national resources for the purpose of re-establishing industries.

renewable energy [rɪ'nju:əbl energy], p. 57
erneuerbare Energie – Energy from → resources which can be replenished, such as sunlight, wind, rain, tides, waves, and geothermal heat.

reprocessing facility [ri:'prəʊsesɪŋ fə'sɪləti], p. 65
Wiederaufbereitungsanlage – Factory to chemically separate and recover energy-rich plutonium from irradiated nuclear → fuel.

reserve [rɪ'zɜ:v], p. 57
Reserve – That part of a → resource that can be recovered economically with the available technologies.

resource [rɪ'sɔ:s], p. 57
Ressource – Total amount of undiscovered and discovered material that exists.

retirement age [rɪ'taɪəmənt eɪdʒ], p. 127
Rentenalter – The age at which most people normally retire from work.

reunification [ˌri:ˌju:nɪfɪ'keɪʃn], p. 151
Wiedervereinigung – The act of joining together two or more regions or parts of a country so that they form a single political unit again.

revitalisation [ˌri:ˌvaɪtəlaɪ'zeɪʃn], pp. 31, 149
Revitalisierung, Wiederbelebung – Process of making something stronger, more active or more healthy again.

run-of-the-river power station [rʌn əv ðə 'rɪvə(r) 'paʊə steɪʃn], p. 67
Laufwasserkraftwerk – A → power station utilising the river flows to generate power.

rural area ['rʊərəl 'eəriə], pp. 111, 129
ländliches Gebiet – Countryside; in contrast to an → urban area.

rural-to-urban migration ['rʊərəl tə 'ɜ:bən maɪ'greɪʃn], p. 53
Land-Stadt Migration – The movement of people from the countryside to city areas.

S

salinisation [sə'lɪnaɪ'zeɪʃ(ə)n], p. 85
Versalzung – The process of increasing the salt content in the soil.

Schengen Agreement ['ʃeŋən əgri:mənt], pp. 107, 109
Schengener Abkommen – A treaty signed on June 14th, 1985 which created the so-called Schengen Area in Europe – an area of 26 European states in which border checks have officially been abolished.

secondary energy ['sekəndri 'enədʒi], p. 57
Sekundärenergie – Energy which is transformed from primary sources like electricity, gasoline or methane.

secondary sector ['sekəndri 'sektə(r)], p. 125
sekundärer Sektor – The portion of an → economy that includes light and heavy industrial → manufacturers of finished goods and products from raw materials.

segregation [ˌsegrɪ'geɪʃn], pp. 27, 31
Segretation, Trennung – Enforced distribution of specific groups of people into distinct areas, usually based on, e.g. race and/or religion.

shopping center ['ʃɒpɪŋ sentə(r)], p. 149
Einkaufszentrum – A modern form of shopping in which one or more buildings form a complex of shops representing merchandisers with interconnecting walkways that enable customers to walk walk everywhere inside.

shrinkage ['ʃrɪŋkɪdʒ], p. 149
Schrumpfen – The decline of a certain factor like population or production.

smart grid [smɑ:t grɪd], p. 77
intelligentes Stromnetz – An electricity network based on digital technology that supplies electricity to consumers via two-way digital communication to help improve efficiency, reduce energy consumption and costs.

soil compaction [sɔɪl kəm'pæktʃn], p. 85
Bodenverdichtung – The increase of bulk density or decrease in porosity of soil due to externally or internally applied loads.

soil conservation [sɔɪl ˌkɒnsə'veɪʃn], p. 97
Bodenschutz – The prevention of soil loss from erosion or reduced fertility caused by overuse, acidification, → salinisation or other chemical soil contamination.

soil degradation [sɔɪl ˌdegrə'deɪʃ(ə)n], p. 85
Bodendegradierung – The process of declining soil productivity and fertility caused by erosion. It may be accelerated by human activities.

soil erosion [sɔɪl ɪ'rəʊʒ(ə)n], pp. 85, 97
Bodenerosion – A naturally occurring process of the displacement of upper layer of soil by water and wind which might be accelerated by human activities.

soil fertility [sɔɪl fəː'tɪlɪti], p. 85
Bodenfruchtbarkeit – The ability of a soil to supply plant growth.

solar cell [ˌsəʊlə 'sel], p. 73
Solarzelle – A cell designed to convert sunlight into electrical energy, mostly consisting of layers of specially prepared silicon.

solar collector [ˌsəʊlə kə'lektə(r)], p. 73
Sonnenkollektor – Device to capture and use solar radiation for heating air or water.

solar energy [ˌsəʊlə 'enədʒi], p. 73
Sonnenenergie – The energy the Earth receives from the sun.

solar panel [ˌsəʊlə 'pænl], p. 73
Solarmodul – An electrical device consisting of a large array of connected → solar cells.

solar tower [ˌsəʊlə 'taʊə(r)], p. 73
Solarturm – A tall tower in which light is gathered from surrounding mirrors to produce → concentrated solar power.

special economic zone (SEZ) ['speʃl ˌiːkə'nɒmɪk zəʊn], pp. 25, 35
Sonderwirtschaftszone – An area with particular economic advantages, for example lower taxes compared with the rest of the country, to encourage investment and development there.

specialisation [ˌspeʃəlaɪ'zeɪʃn], p. 139
Spezialisierung – A method of production where a business, area, or → economy focusses production on a limited variety of products or services to gain greater efficiency.

standard of living ['stændəd əv 'lɪvɪŋ], p. 37
Lebensstandard – A level of material comfort as measured by the goods, services, and luxuries available to an individual, group, or nation.

steel crisis [stiːl 'kraɪsɪs], pp. 141, 143
Stahlkrise – Crisis of the steel industry that started in the 1970s and hit the → old industrial regions of Europe, for example. Its consequences include jobloss and the decline of industrial regions due to a sinking demand for steel and falling prices.

stock exchange ['stɒk ɪkstʃeɪndʒ], pp. 27, 145
Börse – Organised and regulated financial market where stocks, bonds, and other securities are bought and sold governed by the forces of demand and supply.

structural change ['strʌktʃərəl tʃeɪndʒ], pp. 137, 143
Strukturwandel – Change in the socio-economic structure of an area over a longer period of time, e.g. after the decline of a major industry and the introduction of other types of industry.

structural funds ['strʌktʃərəl fʌnds], p. 113
Strukturfonds – Funds created by the → EU to support the economic development in poorer EU member states.

subsidy ['sʌbsədi], p. 115
Subvention, Fördermittel – A benefit for groups, companies or individuals granted by the government; usually in form of a cash payment or a reduction of taxes. Typically, it is given to reduce a certain burden.

suburbanisation [sʌbˌɜːbənaɪ'zeɪʃn], pp. 53, 131, 147, 149
Suburbanisierung – Movement of population and economic activities from within cities to the edges.

sustainability [səˌsteɪnə'bɪləti], pp. 95, 97, 103
Nachhaltigkeit – Balance between economic, social and ecological activities to secure → resources for future generations.

sustainable development [sə'steɪnəbl dɪ'veləpmənt], pp. 95, 101
nachhaltige Entwicklung – Development that meets the needs of the present without compromising the ability of future generations to meet their own needs.

Sustainable Development Goals (SDGs) [sə'steɪnəbl dɪ'veləpmənt goʊls], p. 95
Ziele für nachhaltige Entwicklung – Political targets established by the United Nations in order to secure worldwide → sustainable development economically, socially and ecologically.

sustainable forestry [sə'steɪnəbl 'fɒrɪstri], p. 95
nachhaltige Forstwirtschaft – The environmentally appropriate, socially beneficial, and economically viable management of forests for present and future generations.

T

tanker ['tæŋkə(r)], p. 63
Tankschiff – A ship designed for the transport of liquids or gases.

taxation [tæk'seɪʃn], p. 119
Besteuerung – A charge imposed by the government on the citizens and companies of a country in order to finance the government's expenses.

terms of trade [tɜːmz əv treɪd], p. 21
Handelsverhältnis – Relationship between the prices of → import and → export.

tertiary sector ['tɜːʃəri 'sektə(r)], pp. 111, 125
Tertiärer Sektor – Sector of the economy concerned with the sale and use of economic goods and services.

tidal barrage ['taɪdl 'bærɑːʒ], p. 75
Gezeitensperrmauer – A dam-like structure used to capture the energy from masses of water moving in and out of a bay or river due to tidal forces.

tidal energy ['taɪdl 'enədʒi], p. 75
Gezeitenenergie – Energy obtained by harnessing energy from the shifting water in ocean tides.

topsoil [tɒpsɔɪl], p. 85
Oberboden – The upper, outermost layer of soil, usually the top 2 inches (5.1 cm) to 8 inches (20 cm). It has the highest concentration of organic matter and microorganisms.

transformation [ˌtrænsfə'meɪʃn], p. 129
Transformation – In an organizational context, a process of profound and radical change that orients an organization in a new direction and takes it to an entirely different level of effectiveness.

transnational company [ˌtrænzˈnæʃnəl ˈkʌmpəni], p. 27
transnationales Unternehmen – Company which owns or controls production facilities in more than one country through → foreign direct investment.

Treaty of Maastricht [ˈtriːti əv mɑːstrɪtʃt], p. 109
Maastrichter Vertrag – A treaty signed on February 7th, 1992 which founded the → EU. It became effective on November 1st, 1993 and paved the way for the introduction of the euro.

triad economy [ˈtraɪæd ɪˈkɒnəmi], p. 25
– The Triad refers to the three centres dominating the → world economy until the late 1990's: North America, (Western) Europe and Japan.

turbine [ˈtɜːbaɪn], p. 67
Turbine – Machine in which the kinetic energy of moving water, steam, air, or combustion products of a → fuel is converted into mechanical energy by causing a bladed rotor to rotate.

U

unconventional oil [ˌʌnkənˈvenʃənl ɔɪl], p. 61
Öl aus unkonventionellen Ressourcen – Crude oil that is not produced by traditional extraction methods which includes offshore, oil sands, and tight oil.

undernutrition [ˈʌndənjuˈtrɪʃn], p. 49
Unterernährung – A lack of food or a lack of the right type of food over a period of time which can lead to poor health, a limited capacity to work and/or death by starvation.

unemployment [ˌʌnɪmˈplɔɪmənt], p. 129
Arbeitslosigkeit – 1. The state of people not having a job; 2. The number of people without a job, often as a percentage of the total → labour force.

urban area [ˈɜːbən ˈeəriə], p. 129
städtisches Gebiet – City or a town; in contrast to an → rural area.

urbanisation [ˌɜːbənaɪˈzeɪʃn], pp. 53, 101
Urbanisierung – Expansion of towns in number, area and population.

V

virtual power plant [ˈvɜːtʃuəl ˈpaʊə plɑːnt], p. 77
virtuelles Kraftwerk – A network of decentralized, medium-scale power generating units such as → combined heat and power (CHP) units, → wind farms and solar parks.

virtual water [ˈvɜːtʃuəl ˈwɔːtə(r)], p. 87
virtuelles Wasser – The hidden flow of water if food or other commodities are traded from one place to another.

W

water availability [ˈwɔːtə(r) əˌveɪləˈbɪləti], p. 87
Wasserverfügbarkeit – The probability of water resources.

water distribution [ˈwɔːtə(r) ˌdɪstrɪˈbjuːʃn], p. 87
Wasserverteilung – The dissemination of water on Earth.

water management [ˈwɔːtə(r) ˈmænɪdʒmənt], p. 87
Wasserwirtschaft – All activities of planning, developing, and distributing the optimum use of water resources.

wind energy [ˈwɪnd ˈenədʒi], p. 71
Windkraft – Energy contained in blowing wind.

wind farm [ˈwɪnd fɑːm], p. 71
Windpark – An area of land with a cluster of → wind turbines.

wind lens [ˈwɪnd lenz], p. 71
Windlinse– A modified → wind turbine.

wind turbine [ˈwɪnd ˈtɜːbaɪn], p. 71
Windrad – A tall structure with rotating blades that generates electricity.

workforce [ˈwɜːkfɔːs], p. 131
erwerbstätige Bevölkerung – All the people in a country or an area who are engaged in work.

world economy [wɜːld ɪˈkɒnəmi], p. 125
Weltwirtschaft – The economy of the world, considered as an international exchange of goods and services.

world trade [wɜːld treɪd], p. 11
Welthandel – The international exchange of capital, goods, and services.

world wide web [wɜːld waɪd ˈweb], p. 15
World Wide Web – Part of the internet containing websites and webpages.

Phonetic symbols									
Vowels				Dipthongs		Consonants			
ɑː	arm	i	happy	aɪ	eye	ŋ	song	ð	these
ʌ	but	iː	easy	aʊ	our	r	red	θ	bathroom
e	desk	ɒ	orange	eə	air	s	sister	v	very
ə	an	ɔː	all	eɪ	take	z	zebra	w	what
ɜː	bird	ʊ	look	ɪə	here	ʒ	television	Stress	
æ	apple	u	January	ɔɪ	boy	dʒ	sausage	ˈ	Main stress in a word
ɪ	in	uː	boot	əʊ	go	ʃ	fresh	ˌ	Secondary stress in a word coming before the main stress.
				ʊə	your	tʃ	child		

www.westermann.de/links/114038

Sources

|123RF.com, Hong Kong: skyfish555 130 M3. |action press, Hamburg: NIBOR 88 M1. |Advantage Media Services, Vechta: M. Niehues 139 M6 li.. |akg-images GmbH, Berlin: 96 M3. |alamy images, Abingdon/Oxfordshire: cunard, nick 99 M6; Dinodia Photos 98 M4; Frank Bienewald 99 M5; Ian Dagnall 60 M2; imageBROKER 72 M1, 142 M2; Jim West 31 M5; Joerg Boethling 70 M1; LatitudeStock 77 M4; Michael Dwyer 65 M5; petrut calinescu 110 M1 li.; PhotoStock-Israel 36 M1; Pulsar Images 84 M3; Richard Wareham Fotografie 100 M1; Tim Gainey 68 M1; Universal Images Group North America LLC/DeAgostini 75 M4 Foto; ZUMA Press, Inc. 98 M1. |Andernach.net GmbH, Andernach: © Christoph Maurer 103 M5. |APA-PictureDesk GmbH, Wien: van 't Wout, Erik 51 M4 Foto. |Baaske Cartoons, Müllheim: Jules Stauber 42 M1; K.G. Striepecke 48 M1. |Bilderberg, Hamburg: Rainer Drexel 44 M1 li.. |Cagle Cartoons, Santa Barbara, CA: 82 M2, 102, 154 102 M1 und 154 M1. |CLAAS KGaA mbh, Harsewinkel: 68 M3. |Colourbox.com, Odense: Mats Ljungberg 73 m.. |ddp images GmbH, Hamburg: Latz, M. 124 M1 m.. |Deutsche Post AG, Bonn: 2014 149. |DRÄXLMAIER Group, Vilsbiburg: 17 M3. |dreamstime.com, Brentwood: Bayda127 27 M5; Brett Critchley 36 M3; Minh Tang 5 li.m.; Piccaya 38 M1 o.re.; Vangelis 122; Zaramira 32 M3. |EnBW Energie Baden-Württemberg AG, Karlsruhe: 66 M2. |Eyferth, Konrad, Berlin: 4, 157 4 u. ff, 159. |Flughafen München, München: 146 M3 Foto. |Fotoarchiv Ruhr Museum, Essen: Fotograf unbekannt 140 M1. |fotolia.com, New York: Andrey Popov 14 M1 li.; Carey, Richard 89 M6; Christian Schwier 136 M1; contrastwerkstatt 52 M1, 132 M1 li.; fotoblitzch 132 M1 m.; Fotolyse 137 M6 re.; gemenacom 86 M1 li.; Horst Schmidt 72 M2 re.; inacioluc 103 M3; Jargstorff, Wolfgang 137 M6 li.; Johannsen, M. 6; kameraauge 5 re.; Kara 17 M7; leeyiutung 22 o.; lumen-digital 4, 134 4 m.re. und 134 M1; moonrun 106; Nivens, Sergey 8 /9 Hintergrund; Noppasinw 29 M6; Robert Schneider 84 M1; rotschwarzdesign 131 M4; salman2 17 M5; Sanders, Gina 139 M6 re.; SyB 4 o.li., 38; Voyant 52 M2; vschlichting 124 M1 li.; Yakusheva, Alena 44 M1 re.. |Fraport AG, Frankfurt/Main: Stefan Rebscher 145 M4. |Gesellschaft für ökologische Forschung e.V., München: Sammlung 81 M5; Sammlung / Wolfgang Zängl 81 M6. |Getty Images, München: AFP 31 M6 o., 120 M4 u.; Bettmann 89 M4; Essick, Peter 78; Rodrigo Arangua/AFP 12 M1; Steve Jennings/WireImage 11 M4. |Global Footprint Network/www.footprintnetwork.org, Oakland: 90 M1. |Güttler, Peter - Freier Redaktions-Dienst (GEO), Berlin: 90 M2. |Haitzinger, Horst, München: 95 M6, 126 M1. |Huber Images, Garmisch-Partenkirchen: 146 M1. |i.motion gmbh, Bamberg: 58 M1, 60 M1, 62 M1, 75 M3, 76 M1; Zimmermann, Uli 16 M1. |Imaginechina, Shanghai: Wu kanjun 34 M1. |Imago, Berlin: Oed, Hans-Günther 124 M1 re.. |iStockphoto.com, Calgary: AdrianHancu 5 li.u.; ake1150sb 28 M1; anek_s 34 M3; bennymarty 18 M1; borgogniels 36 M2 o.; canaran 20 M1; ChrisSteer 5 li.o.; David Sucsy 97 M6 li.; DesignerforU 26 M1; DutchlightNetherlands 94 M1; elxeneize 151 M6; imagixian 31 M6 u.; JackF 36 M2 u.; mapodile 16 M2; Pavone, Sean 144 M1; skhoward 25 M4. |Jüngst, Reiner, Wolfenbüttel: 12 M4. |Karto-Grafik Heidolph, Dachau: 82, 133, 140, 142. |laif, Köln: Alex MacLean/NYT/Redux 30 M1. |mauritius images GmbH, Mittenwald: Reiner Elsen 26 M4, 148 M3. |Möller Industrietechnik GmbH, Hecklingen: 113 M3 li.; Lutz Krüger 113 M3 re.. |Morland, Morten: 121 M5. |NASA, Washington: 3, 3; Erde 80. |NASA - Earth Observatory: 178 li.o., 178 li.u., 178 re.u., 179 li.o., 179 li.u., 179 re.o., 179 re.u.; Adam Voiland und Jeff Schmaltz 101 M5; Visible Earth 178 re.o.. |NASA/GSFC, Houston/Texas: Visible Earth 178 /179. |Nordstern GmbH, Gelsenkirchen: 141 M6 Foto. |PantherMedia GmbH (panthermedia.net), München: Angelika Bentin 35 u.re.; Heiber, Werner 4 o.re. und; TReinhard 54 o.. |photothek.net GbR, Radevormwald: Liesa Johannssen 72 M2 li.. |Picture-Alliance GmbH, Frankfurt/M.: AP/Vogel 18 M3; blickwinkel/P. Schuetz 140 M3; ChinaFotoPress/Shepherd Zhou/MAXPPP 10 M2; dpa 144 M2; dpa-Zentralbild/euroluftbild.de/robert Grahn 145 M5; dpa/T. Brakemeier 120 M1; European Commission/Georges Boulougouris/Ec/Handout 120 M4 o.; PictureLux/Legacy 96 M1; PIXSELL 104; Tödt, Matthias 132 M1 re.; Westend61 / Maartje van Caspel 14 M1 m.; ZB 149 M4; ZB/euroluftbild.de/Hans Blossey 142 M1; ZB/Grubitzsch, W. 148 M1. |Plaßmann, Thomas, Essen: 39 o.. |Raupach, Thomas, Hamburg: 54 u.. |Scott Hydroelectric: 66 M4. |Shutterstock.com, New York: Andrew Zarivny 66 M1; Atelier211 8 M2; Dragon Images 52 M3; Earl D. Walker 97 M6 re.; Fotogien 33 M6; Frank11 110 M1 re.; Gerhard Roethlinger 8 M1; kamui29 9 M3; Marchan 86 M1 re.; Mukesh Kumar Jwala 14; RomanSlavik.com Titel; Sorbis 9 M4; Syda Productions 18 M4; ValeStock 40. |Spedition Schmechtig Schwerlast GmbH /www.Spedition-Schmechtig.de, Genthin: 70 M4. |Stadt Sonneberg: 129 M5. |Tekülve, Rita, Essen: 138 M3. |thyssenkrupp Steel Europe AG, Duisburg: thyssenkrupp Konzernarchiv, Duisburg 143 M5 o.re.. |ullstein bild, Berlin: Engelke 151 M4 re.; Hampel 151 M4 li.. |United Nations - UNRIC Verbindungsbüro in Deutschland, Bonn: 95. |United Nations Photo Library/Department of Public Information, New York: Devra Berkowitz 27 M6. |Visum Foto GmbH, München: Crispin Hughes/Panos Pictures 22 u.. |Webb Aviation, Cheshire: 92. |wikimedia.commons: Baycrest/Lizenz: cc by-sa-2.5 35 o.re.; CellarDoor85/Lizenz: CC BY-SA-3.0 116 M3; James Smythe, CC-BY-SA-3.0 35 o.li.; Mike Bonzalez/Lizenz: CC BY-SA-3.0 74 M1. |Willhardt, Jens, München: 147 M6. |www.avda-foto.de: Lizenz: CC BY-SA-3.0 150 M1. |www.worldmapper.org, London: 4 m.li., 24 M1.

Wir arbeiten sehr sorgfältig daran, für alle verwendeten Abbildungen die Rechteinhaberinnen und Rechteinhaber zu ermitteln. Sollte uns dies im Einzelfall nicht vollständig gelungen sein, werden berechtigte Ansprüche selbstverständlich im Rahmen der üblichen Vereinbarungen abgegolten.

Different Types of Numbers

1. Reading and writing numbers

.	**point** (for decimals, e.g. 2.54). See decimals.
,	Use a comma or a blank for groups of three numbers. (Do not read commas.) See below.
0	**zero**, "0", nought (UK)
1–10	In a text write **one** to **ten** as words.
>10	In a text write numbers **larger than** 10 as numbers, e.g. "56", not "fifty-six".
123	a/one **hundred** (and) twenty-three
234	two **hundred** (and) thirty-four
5,678 (5 678)	five **thousand** six hundred (and) seventy-eight
93,000,000	ninety-three **million**
6,000,000,000	six **billion** (milliard = UK old form)

2. Different types of numbers

Ordinal numbers

1st	first	7th	seventh
2nd	second	8th	eighth
3rd	third	9th	ninth
4th	fourth	10th	tenth
5th	fifth	11th	eleventh
6th	sixth	12th	twelfth

Fractions

Fraction	We say ...	Decimal	Per cent
1/2	a/one half (of) / one out of two	0.5	50 %
1/3	a/one third (of) / one out of three	0.33	33 ⅓ %
1/4	a/one fourth (of) / one out of four	0.25	25 %
1/5	a/one fifth (of) / one out of five	0.2	20 %
1/10	a/one tenth (of) / one out of ten	0.1	10 %
1/20	a/one twentieth (of) / one out of twenty	0.05	5 %
1/100	a/one hundredth (of) / one out of a/one hundred	0.01	1 %

Decimals

0.5	(zero) point five
3.14 (π)	three point one four (pi)

Powers / Exponents

10^2	ten **squared** ten **to the power** of two ten **to the** second	**100**
10^3	ten **to the power of** three ten **to the** third	**1000**

The Earth is 93 x 10^6 miles from the sun.
(ninety-three times ten to the sixth) – 93 million –

3. Arithmetic

= Proportions

=	is, is **equal to, equals**	≠	is **unequal to**, is **not equal to**
>	is **more**, is **larger than**	<	is **less**, is **fewer**, is **smaller than**
4:1	four **to** one	4 x	four **times as large as**

+ Addition → Sum

12	+	4	=	16
Twelve	**plus**	four	equals	sixteen.

– Subtraction → Difference

12	–	4	=	8
Twelve	**minus**	four	equals	eight.

X Multiplication → Product

12	x	4	=	48
Twelve	**times, multiplied by**	four	equals	forty-eight.

: Division → Quotient

12	: /	4	=	3
Twelve	**divided by**	four	equals	three.

4. Geometry

parallel lines

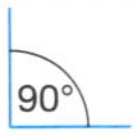

right angle = 90°
A right angle has got ninety **degrees**.

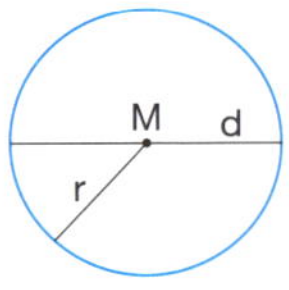

circle
A circle is **round** or **circular**. It has got a **middle point (M)**. The line from the middle point to the circle wall is the **radius (r)**. The line from one side of the circle through the middle point to the other side is the **diameter (d)**.

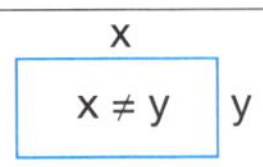

rectangle
A rectangle is **rectangular**. It has got two parallel equal sides on two sides at right angles.

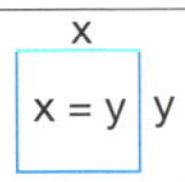

square
A square is **square**. It has got four equal parallel sides at right angles.

triangle
A triangle is **triangular**. It has got three sides.

3-d figures

cube

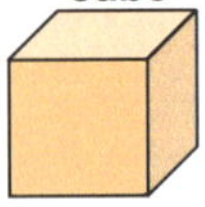

sphere

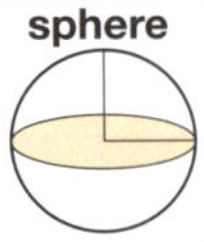

5. Time

1 minute	= 60 seconds	**2006**
60 minutes	= 1 hour	**two thousand (and) six**
24 hours	= 1 day	**twenty "o" six**
365 ¼ days	= 1 year	
100 years	= 1 century	